Production and Efficiency Analysis with R

Andreas Behr

Production and Efficiency Analysis with R

Springer

Andreas Behr
Faculty of Business Administration
and Economics
University of Duisburg-Essen
Essen, Germany

ISBN 978-3-319-20501-4 ISBN 978-3-319-20502-1 (eBook)
DOI 10.1007/978-3-319-20502-1

Library of Congress Control Number: 2015960276

Mathematics Subject Classification (2010): 90C05, 65K10

Springer Cham Heidelberg New York Dordrecht London

Printed on acid-free paper

Springer-Verlag GmbH Berlin Heidelberg is part of Springer Science+Business Media (www.springer.com)

Preface

The book is based on my lecture notes written for the course in production and efficiency analysis at the Münster University and University of Duisburg-Essen. Some basic knowledge of statistical methods, regression analysis, and the R programming environment would be helpful for understanding the text.

I am grateful to many students of my courses for their feedback. I would like to thank Christoph Schiwy for providing the layout of the manuscript using Latex and knitr and Katja Theune, Anastasia Diel, and Jurij Weinblat for reading the manuscript.

Münster, Germany
June 2015

Andreas Behr

Contents

1 Introduction 1
1.1 A Short Introduction to R 1
1.1.1 Objects 1
1.1.2 Dataframes 2
1.1.3 Sequences, Replications, and Loops 3
1.1.4 Matrices 4
1.1.5 Reading and Storing Data Files 6
1.1.6 Probability Distributions 7
1.1.7 Graphics 8
1.1.8 Linear Regression 9
1.2 Testing Linear Restrictions 10
1.3 Maximum Likelihood Estimation 15
1.3.1 The Basic Idea of Maximum Likelihood Estimation 15
1.3.2 The Exponential Distribution 15
1.3.3 Properties of ml-Estimators 18
1.4 Numerical Optimization 22
1.4.1 The General Form of Algorithms 22
1.4.2 The Newton–Raphson Algorithm 23
1.4.3 Two Simple Examples 24
1.5 Recommended Readings 30
1.6 Exercises 30
References 32

2 Linear Production Model 33
2.1 The Theoretical Linear Production Model 33
2.1.1 Industries and Goods 33
2.1.2 Direct and Indirect Inputs 35
2.1.3 The Employment Model 36
2.1.4 A Numerical Example with R 38
2.2 Input–Output Tables 41
2.2.1 Aggregated Tables for Germany 41
2.2.2 Some Identities 42

2.3 Analyzing German Input–Output Tables 1991 and 2007 44
2.3.1 Handling Input–Output Tables 44
2.3.2 Some Empirical Results 46
2.4 Recommended Readings 55
2.5 Exercises 56
References 56

3 Production Functions 57
3.1 Introduction 57
3.2 The Cobb–Douglas Function 59
3.2.1 Properties of the Function 59
3.2.2 Estimating the CD with R 60
3.3 The Constant Elasticity of Substitution Function 62
3.3.1 Properties of the Function 63
3.3.2 Estimating the CES with R 66
3.4 The Transcendental Logarithmic Production Function 71
3.4.1 Properties of the Function 72
3.4.2 Estimating the TL with R 75
3.5 Recommended Readings 79
3.6 Exercises 80
References 81

4 Production Functions with Panel Data 83
4.1 Introduction 83
4.2 Introduction to Panel Data 84
4.2.1 Unrestricted and Restricted Models 84
4.2.2 Pooled, Within, Between 85
4.2.3 Random Effects 87
4.2.4 Fixed or Random? 94
4.3 Panel Data Analysis with R 95
4.3.1 Generating a Panel Data Set 95
4.3.2 Some Transformations with Panel Data 98
4.3.3 Estimating Panel Data Models Using the plm Package 99
4.3.4 Dynamic Panel Data Models 104
4.4 Recommended Readings 108
4.5 Exercises 109
References 111

5 Linear Programming 113
5.1 Introduction 113
5.1.1 Cost Minimization 113
5.1.2 Profit Maximization 116
5.1.3 General Notation 117
5.1.4 Convex Sets 118
5.1.5 Slacks and Surpluses 119

5.2 Simplex Algorithm 120
5.2.1 Maximization 120
5.2.2 Minimization 122
5.2.3 Duality 125
5.2.4 Simplex Tableau 127
5.3 Solving the Simplex Using R 129
5.3.1 The Setup of Simplex Tableaus 129
5.3.2 Solving Some Examples 129
5.4 Recommended Readings 132
5.5 Exercises 133
References 134

6 Data Envelopment Analysis 135
6.1 Introduction 135
6.1.1 Productivity and Efficiency 135
6.1.2 One Input–One Output 136
6.1.3 Two Inputs–One Output 137
6.1.4 One Input–Two Outputs 138
6.1.5 Fixed and Variable Weights 140
6.2 Charnes–Cooper–Rhodes-Model 141
6.2.1 Introduction 141
6.2.2 Programming Problem 142
6.2.3 Examples 143
6.2.4 Production Correspondence 144
6.2.5 Output-Oriented Model 148
6.2.6 Returns to Scale in DEA Models 149
6.3 Data Envelopment Analysis with R 151
6.3.1 Benchmarking Package 151
6.3.2 Examples 151
6.4 Recommended Reading 157
6.5 Exercises 158
References 159

7 Stochastic Data Envelopment Analysis 161
7.1 Introduction 161
7.2 Population and Simple Random Samples 162
7.3 The Bootstrap 164
7.4 Sampling and DEA 167
7.5 DEA and the Naive Bootstrap 171
7.5.1 Some Notation 171
7.5.2 Bootstrap Estimation Based on a Specific Sample 172
7.5.3 Bootstrap Estimation Based on All Samples 175
7.6 The Simar–Wilson Approach 176
7.6.1 Strongly Simplified Algorithm 176
7.6.2 The Simar–Wilson Algorithm 178

7.7 Recommended Reading 181
7.8 Exercises 181
References 182

8 Stochastic Frontier Analysis 183
8.1 Introduction 183
8.2 Production and Inefficiency 183
8.2.1 Observed and Efficient Production 184
8.2.2 Production Frontier and Deviations 184
8.2.3 Data Generation 185
8.2.4 Corrected Ordinary Least Squares 189
8.2.5 Corrected Ordinary Least Squares Using R 189
8.3 Maximum Likelihood Estimation 191
8.3.1 The Log-Likelihood 192
8.3.2 Estimation of Individual Inefficiency Terms 193
8.3.3 Maximum Likelihood with R 195
8.4 Recommended Reading 199
8.5 Exercises 200
References 201

9 Panel Data Stochastic Frontier Analysis 203
9.1 Introduction 203
9.2 Homogeneity and Firm Specific Time Invariant Inefficiency 204
9.3 Time Varying Firm Specific Inefficiency 208
9.4 The Stochastic Frontier Model with Time Varying Inefficiency 212
9.4.1 Sketching the Idea of the Approach 213
9.4.2 Obtaining the Log-Likelihood 214
9.4.3 Generating Data According to the Model 217
9.5 Recommended Reading 221
9.6 Exercise 221
References 222

Functions Index 223

Subject Index 225

1 Introduction

Abstract

In this chapter, we provide a very concise introduction to R. Furthermore, we briefly introduce some statistical concepts: Testing linear restrictions in regression models, the method of maximum likelihood estimation, and methods of numerical optimization.

1.1 A Short Introduction to R

1.1.1 Objects

R is an object oriented programming environment for data manipulation, calculation, and graphical display. Objects (e.g., vectors, matrices, tables, and so on) can be given some content using the assignment symbol `<-`. Functions can be called with arguments given in brackets. For example `c()` combines some elements which we name `a`:

```
a <- c(1,2)
```

We can print the object simply by calling its name, apply functions to it, and do some arithmetics:

```
a

## [1] 1 2

mean(a)

## [1] 1.5

a+1

## [1] 2 3
```


A. Behr, *Production and Efficiency Analysis with R*,
DOI 10.1007/978-3-319-20502-1_1

The in-built help-function can be called using a question mark:

```
?mean
```

1.1.2 Dataframes

Data are stored in a dataframe which contains the values and some attributes. As a simple example, we create a small dataframe d using the command `data.frame`:

```
d <- data.frame(x = c("A","B","C"),
                y = c(22,35,41),
                z = c(0,1,0))
d

##   x  y z
## 1 A 22 0
## 2 B 35 1
## 3 C 41 0
```

Variables of the dataframe d can be selected either by using $ or by using indices:

```
d$x

## [1] A B C
## Levels: A B C

d[1]

##   x
## 1 A
## 2 B
## 3 C
```

Individual elements (or parts of the dataframe) can be selected using square brackets []:

```
d[1, 2]

## [1] 22

d[2:3,1:2]

##   x  y
## 2 B 35
## 3 C 41

d[,2]

## [1] 22 35 41

d[2,]

##   x  y z
## 2 B 35 1
```

1.1.3 Sequences, Replications, and Loops

Sequences can be generated using the command `seq(from,to,by)`.

```
seq(0.8,0.9,0.02)
```

```
## [1] 0.80 0.82 0.84 0.86 0.88 0.90
```

A short form for integers and step size 1 is:

```
1:5
```

```
## [1] 1 2 3 4 5
```

Replications can be generated by

```
rep(1:2,times=3)
```

```
## [1] 1 2 1 2 1 2
```

```
rep(1:2,each=3)
```

```
## [1] 1 1 1 2 2 2
```

```
rep(1:2,times=2:1)
```

```
## [1] 1 1 2
```

Loops are often used for control structures. An example is the following `for`-loop:

```
for (x in 3:5) {print(paste("The value of x is",x))}
```

```
## [1] "The value of x is 3"
## [1] "The value of x is 4"
## [1] "The value of x is 5"
```

In many cases, we can avoid loops by using the functions `apply` or `tapply`. Using `apply(X, MARGIN, FUN)`, we apply the function specified with the argument FUN towards rows (MARGIN=1) or columns (MARGIN=2) of X. To demonstrate `apply`, we construct a matrix and calculate the sums of the elements in each column:

```
x <- matrix(1:9,ncol=3,byrow=F)
x
```

```
##      [,1] [,2] [,3]
## [1,]    1    4    7
## [2,]    2    5    8
## [3,]    3    6    9
```

```
apply(X=x,MARGIN=2,FUN=sum)
```

```
## [1]  6 15 24
```

An abbreviation making use of the ordering of the commands is:

```
apply(x, 2, sum)
```

```
## [1]  6 15 24
```

Note that we can define a function which we apply to columns or rows of X. For example we may want to select the a-th smallest element of a vector:

```
mysort <- function(y,a) sort(y)[a]
mysort(c(3,7,9,6),2)
```

```
## [1] 6
```

Choosing $a = 2$, we apply the function towards the columns of X (here: $X = y$):

```
y <- matrix(rnorm(9),ncol=3,byrow=F)
y
```

```
##         [,1]    [,2]    [,3]
## [1,]  0.2975 -1.4664  0.2488
## [2,]  0.3527  0.8105  0.6948
## [3,] -0.6467  0.1442 -0.6412
```

```
apply(y, 2, mysort, 2)
```

```
## [1] 0.2975 0.1442 0.2488
```

The function `tapply(X, INDEX, FUN)` applies a function specified with FUN towards the elements in groups of X, which are specified using `INDEX`.

```
x <- 1:20
i <- rep(1:5,4)
tapply(x,i,mean)
```

```
##    1    2    3    4    5
##  8.5  9.5 10.5 11.5 12.5
```

1.1.4 Matrices

Contrary to numeric vectors, matrices have a dimension, that is a number of rows and columns:

```
y <- 1:3
dim(y)
```

```
## NULL
```

```
x <- as.matrix(y)
x
```

```
##      [,1]
## [1,]    1
```

```
## [2,]    2
## [3,]    3
```

```
dim(x)
```

```
## [1] 3 1
```

```
x <- matrix(1:3)
dim(x)
```

```
## [1] 3 1
```

Multiplication applied to vectors is element-wise:

```
1:2
```

```
## [1] 1 2
```

```
3:4
```

```
## [1] 3 4
```

```
1:2*3:4
```

```
## [1] 3 8
```

For matrix multiplication, we have to use the symbol `%*%`:

```
matrix(1:2, ncol = 1)%*%matrix(3:4, ncol = 2)
```

```
##      [,1] [,2]
## [1,]    3    4
## [2,]    6    8
```

We obtain the transpose of a matrix using `t()`. The inverse of a matrix solves the equation $A^{-1} \cdot A = I$:

```
A <- matrix(1:4,nrow = 2)
A
```

```
##      [,1] [,2]
## [1,]    1    3
## [2,]    2    4
```

```
solve(A)
```

```
##      [,1] [,2]
## [1,]   -2  1.5
## [2,]    1 -0.5
```

To generate an identity matrix , we use `diag()` and define the number of rows and columns:

```
diag(2)

##      [,1] [,2]
## [1,]    1    0
## [2,]    0    1
```

Using `diag()`, we can also extract diagonal elements of a matrix:

```
diag(matrix(1:4,nrow = 2))

## [1] 1 4
```

and generate a diagonal matrix with specified elements on its diagonal:

```
diag(1:2)

##      [,1] [,2]
## [1,]    1    0
## [2,]    0    2
```

1.1.5 Reading and Storing Data Files

We save the dataframe `d` as file *test.csv* using the function `write.table` and read it into the workspace using `read.table`[1]:

```
d <- data.frame(x = c("A","B","C"),
                y = c(22,35,41),
                z = c(0,1,0))
write.table(d,"test.csv")
d <- read.table("test.csv")
d

##   x  y z
## 1 A 22 0
## 2 B 35 1
## 3 C 41 0
```

One can also use the specification that is compatible with German defaults in Excel:

```
write.csv2(d,"test.csv",row.names=F)
read.csv2("test.csv")

##   x  y z
## 1 A 22 0
## 2 B 35 1
## 3 C 41 0
```

[1]Note that the default delimiter is a space; hence, the file is actually not a comma separated value (csv) file.

1.1.6 Probability Distributions

Many probability distributions are implemented in R, either in the base library or in special packages. Note that probability distributions describe specific random experiments or processes. Empirical data resulting from observed economic situations will never follow exactly theoretical distributions. Nevertheless, sometimes distributions (e.g., histograms or density estimates) of empirical data look rather similar to theoretical distributions. Using probability distributions with R necessitates the specification of the distribution (e.g., normal) and the function (e.g., quantile function):

1. **Distribution**

   - **norm** normal distribution
   - **unif** uniform distribution
   - **binom** binomial distribution
   - **pois** poisson distribution
   - ...

2. **Function**

   - **d** density function
   - **p** probability function
   - **q** quantile function
   - **r** random number

For the standard normal distribution we obtain e.g.:

```
dnorm(0)

## [1] 0.3989

qnorm(0.975)

## [1] 1.96

pnorm(1.96)

## [1] 0.975

rnorm(3)

## [1]  1.4836  1.3018 -0.3861
```

Note that for reproducibility of random numbers, one has to initialize the generator using `set.seed()`:

```
rnorm(3)

## [1] -0.5387  1.6078 -0.9964
```

```
rnorm(3)

## [1] -0.2099  0.2146  0.9011

set.seed(1)
rnorm(3)

## [1] -0.6265  0.1836 -0.8356

set.seed(1)
rnorm(3)

## [1] -0.6265  0.1836 -0.8356
```

1.1.7 Graphics

R allows very flexible configurations of graphics. Due to the recognition of object classes and default settings, standard graphics can be produced very easily. We give two examples, a histogram and a scatter plot with the regression line according to the method of least squares added (see Fig. 1.1).

```
n <- 1000
x <- rchisq(n,5)
# histogram
hist(x)
u <- rnorm(n,sd=3)
y <- 10+0.5*x+u
# scatterplot
plot(x,y,pch=19,col="darkgrey",cex=0.5)
abline(lm(y~x),lwd=2) # add regression line
```

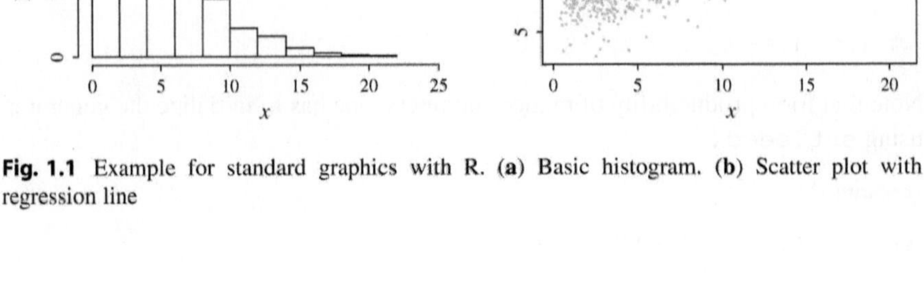

Fig. 1.1 Example for standard graphics with R. **(a)** Basic histogram. **(b)** Scatter plot with regression line

One can export the plots as pdf-files for inclusion in documents using the following command:

```
dev.copy(device=pdf, file="graphic.pdf",
         width=7, height=4, pointsize=10,
         onefile=FALSE, paper="special")
dev.off()
```

1.1.8 Linear Regression

We already made use of the `lm()`-function for estimating linear models in the previous graphic (see Fig. 1.1b). The function `lm()` returns a lm-object with detailed information on the estimated linear model:

```
reg <- lm(y~x)
names(reg)

##  [1] "coefficients"  "residuals"     "effects"
##  [4] "rank"          "fitted.values" "assign"
##  [7] "qr"            "df.residual"   "xlevels"
## [10] "call"          "terms"         "model"
```

For example we obtain the regression coefficients by referring to the sub-object or by calling the function `coef()`:

```
reg$coef

## (Intercept)           x
##      10.060       0.495

coef(reg)

## (Intercept)           x
##      10.060       0.495
```

The command `summary()` provides detailed information on the regression:

```
sreg <- summary(reg)
sreg

##
## Call:
## lm(formula = y ~ x)
##
## Residuals:
##    Min     1Q Median     3Q    Max
## -9.422 -1.923 -0.096  1.944  8.994
##
## Coefficients:
##             Estimate Std. Error t value Pr(>|t|)
## (Intercept)  10.0598     0.1721    58.4   <2e-16 ***
```

```
## x           0.4950     0.0282    17.6    <2e-16 ***
## ---
## Signif. codes:
## 0 '***' 0.001 '**' 0.01 '*' 0.05 '.' 0.1 ' ' 1
##
## Residual standard error: 2.93 on 998 degrees of freedom
## Multiple R-squared:  0.236,       Adjusted R-squared:  0.236
## F-statistic:  309 on 1 and 998 DF,  p-value: <2e-16
```

The summary-object again contains many information on the regression. Using `names()`, we obtain the names of the sub-objects contained in the summary-object:

```
names(sreg)

##  [1] "call"          "terms"         "residuals"
##  [4] "coefficients"  "aliased"       "sigma"
##  [7] "df"            "r.squared"     "adj.r.squared"
## [10] "fstatistic"    "cov.unscaled"
```

For example the coefficient of determination

```
sreg$r.squared

## [1] 0.2363
```

or the variance–covariance matrix of the regression parameters (scaled: $\hat{\sigma}^2 \cdot (X'X)^{-1}$):

```
vcov(reg)

##              (Intercept)          x
## (Intercept)     0.029629 -0.0040892
## x              -0.004089  0.0007935
```

1.2 Testing Linear Restrictions

We start with a simple linear model in matrix notation[2]

$$y = x\beta + u \tag{1.1}$$

with k regressors and n observations. $\beta' = (\beta_0, \ldots, \beta_{k-1})$ is unknown and estimated as

$$\hat{\beta} = (X'X)^{-1}X'y. \tag{1.2}$$

[2]We follow closely the exposition in Johnston and DiNardo, Econometric Methods, 4th edition, page 90–99.

The covariance matrix Σ is unknown and estimated as

$$\hat{\Sigma} = (X'X)^{-1}\hat{\sigma}^2, \qquad \hat{\sigma}^2 = \frac{1}{n-k}\hat{u}'\hat{u}, \qquad \hat{u} = y - X\hat{\beta}. \tag{1.3}$$

The general setup of hypothesis testing is of the form

$$R\beta = r \tag{1.4}$$

where R is a $q \times k$ matrix of constants and r is a q-vector of constants. For example if we assume $k = 5$ and the hypothesis to be tested $\beta_2 + \beta_3 = 1$, we would express this hypothesis as

$$\begin{aligned} R\beta &= r \\ \begin{bmatrix} 0 & 0 & 1 & 1 & 0 \end{bmatrix} \beta &= 1. \end{aligned} \tag{1.5}$$

Assume again $k = 5$ and that we want to test the combined hypothesis $\beta_2 = \beta_3$ and $\beta_1 = 1$. This hypotheses we would translate towards

$$\begin{aligned} R\beta &= r \\ \begin{bmatrix} 0 & 0 & 1 & -1 & 0 \\ 0 & 1 & 0 & 0 & 0 \end{bmatrix} \beta &= \begin{bmatrix} 0 \\ 1 \end{bmatrix}. \end{aligned} \tag{1.6}$$

For testing, we formulate $\text{H}_0\colon R\beta - r = 0$. Intuitively, the test is based on $R\hat{\beta} - r$ and the more $R\hat{\beta} - r$ deviates from 0, the less likely the hypothesis holds. Note that under model assumptions and normality of error terms, we have

$$\begin{aligned} \text{E}(\hat{\beta}) &= \beta & &\longrightarrow & \text{E}(R\hat{\beta}) &= R\beta \qquad \text{and} \\ \text{Var}(\hat{\beta}) &= (X'X)^{-1}\sigma^2 & &\longrightarrow & \text{Var}(R\hat{\beta} - r) &= R(X'X)^{-1}R'\sigma^2. \end{aligned} \tag{1.7}$$

r is constant and $R\hat{\beta}$ is a linear combination of normally distributed random variates, hence, distributed normally. The sum of q normalized (through the inverse covariance matrix) squared normal random variates follows the χ^2-distribution with q degrees of freedom

$$(R\hat{\beta} - r)' \left[R(X'X)^{-1}R'\sigma^2\right]^{-1} (R\hat{\beta} - r) \sim \chi^2(q). \tag{1.8}$$

Because of

$$\begin{aligned}\hat{u} &= y - X\hat{\beta} = y - X(X'X)^{-1}X'y = \left[I - X(X'X)^{-1}X'\right]y \\ &= \left[I - X(X'X)^{-1}X'\right](X\beta + u) \\ &= \left[I - X(X'X)^{-1}X'\right]X\beta + \left[I - X(X'X)^{-1}X'\right]u \\ &= \left[I - X(X'X)^{-1}X'\right]u\end{aligned} \tag{1.9}$$

the estimated residuals are linear combinations of normal random variates. Therefore,

$$\begin{aligned}\hat{u}'\hat{u} &= u'\left[I - X(X'X)^{-1}X'\right]'\left[I - X(X'X)^{-1}X'\right]u \\ &= u'\left[I - X(X'X)^{-1}X'\right]u\end{aligned} \tag{1.10}$$

is a sum of squared normal random variates. Normalizing using the variance of u and r being the rank of $\left[I - X(X'X)^{-1}X'\right]$ we find

$$\frac{\hat{u}'\hat{u}}{\sigma^2} \sim \chi^2(r). \tag{1.11}$$

Finally, we use the ratio of the two independently χ^2-distributed statistics (σ^2 cancels out) to find

$$\frac{\frac{1}{q}(R\hat{\beta} - r)'\left[R(X'X)^{-1}R'\right]^{-1}(R\hat{\beta} - r)}{\frac{1}{n-k}\hat{u}'\hat{u}} \sim \mathrm{F}(q, n-k) \quad \text{or equivalently}$$

$$\frac{1}{q}(R\hat{\beta} - r)'\left[\hat{\sigma}^2 R(X'X)^{-1}R'\right]^{-1}(R\hat{\beta} - r) \sim \mathrm{F}(q, n-k). \tag{1.12}$$

We look at a simple numerical example

$$y = \begin{bmatrix} 3 \\ 1 \\ 8 \\ 3 \\ 5 \end{bmatrix} \quad X = \begin{bmatrix} 1\,3\,5 \\ 1\,1\,4 \\ 1\,5\,6 \\ 1\,2\,4 \\ 1\,4\,6 \end{bmatrix} \tag{1.13}$$

and consider the linear model

$$y = X\beta + u. \tag{1.14}$$

We obtain $\hat{\beta}$ as

$$\hat{\beta} = (X'X)^{-1}X'y = \begin{bmatrix} 4 \\ 2.5 \\ -1.5 \end{bmatrix}. \tag{1.15}$$

In R this example is

```
y <- matrix(c(3,1,8,3,5))
X <- cbind(1,c(3,1,5,2,4),c(5,4,6,4,6))
x <- X[,2:3]
n <- nrow(y)
k <- ncol(X)
reg <- lm(y~x)
sreg <- summary(reg)
b <- matrix(coef(reg))
b

##      [,1]
## [1,]  4.0
## [2,]  2.5
## [3,] -1.5

s2 <- sreg$sigma^2
s2

## [1] 0.75

v <- vcov(reg)
v

##             (Intercept)     x1     x2
## (Intercept)      20.025  3.375 -6.000
## x1                3.375  0.750 -1.125
## x2               -6.000 -1.125  1.875
```

The combined hypothesis we want to test is $\beta_1 = 2$ and $\beta_2 = -1$, now $q = 2$. $R\hat{\beta}$ picks out two elements of $\hat{\beta}$ and $\hat{\sigma}^2 R(X'X)^{-1}R'$ picks out the variances and the covariance. The test statistic is a linear combination of the differences of the two parameters and the constants with variances and covariance as weights

$$\begin{aligned} &\frac{1}{q}(R\hat{\beta} - r)' \left[\hat{\sigma}^2 R(X'X)^{-1}R'\right]^{-1} (R\hat{\beta} - r) \\ &= \frac{1}{2}\begin{bmatrix} 0.5 & -0.5 \end{bmatrix} \begin{bmatrix} 13.33 & 8 \\ 8 & 5.33 \end{bmatrix} \begin{bmatrix} 0.5 \\ -0.5 \end{bmatrix} = 0.33. \end{aligned} \tag{1.16}$$

The p-value is then given by

$$\text{p-value} = 1 - F(0.33, 2, 2) = 0.75. \tag{1.17}$$

In R we obtain the result by the following commands

```
R <- rbind(c(0,1,0),c(0,0,1))
R

##      [,1] [,2] [,3]
## [1,]    0    1    0
## [2,]    0    0    1

r <- matrix(c(2,-1))
q <- nrow(R)
Fs <- 1/q*t(R%*%b-r)%*%solve(R%*%v%*%t(R))%*%(R%*%b-r)
Fs

##        [,1]
## [1,] 0.3333

1-pf(Fs,q,n-k)

##      [,1]
## [1,] 0.75
```

We can test the combined hypothesis $\beta_0 + \beta_1 + \beta_2 = 2$ and $\beta_2 = -1$ as follows:

```
R <- rbind(c(1,1,1),c(0,0,1))
R

##      [,1] [,2] [,3]
## [1,]    1    1    1
## [2,]    0    0    1

r <- matrix(c(2,-1))
q <- nrow(R)
Fs <- 1/q*t(R%*%b-r)%*%solve(R%*%v%*%t(R))%*%(R%*%b-r)
Fs

##       [,1]
## [1,] 2.911

1-pf(Fs,q,n-k)

##        [,1]
## [1,] 0.2557
```

A convenient way to carry out hypothesis tests is using the command `linearHypothesis()` implemented in the car package which leads to identical results:

```
library(car)
linearHypothesis(model=reg,hypothesis.matrix=R, rhs=r)

## Linear hypothesis test
##
## Hypothesis:
## (Intercept)  + x1  + x2 = 2
## x2 = - 1
```

```
##
## Model 1: restricted model
## Model 2: y ~ x
##
##   Res.Df  RSS Df Sum of Sq    F Pr(>F)
## 1      4 5.87
## 2      2 1.50  2      4.37 2.91   0.26
```

1.3 Maximum Likelihood Estimation

1.3.1 The Basic Idea of Maximum Likelihood Estimation

The idea of maximum likelihood (ml) has been introduced by Ronald Fisher (1890–1962) and is very intuitive: choose as an estimate for the unknown parameters $\theta_1, \ldots, \theta_r$ the numerical values, that give the observed values $x_1, \ldots, x_n$ the highest probability. The likelihood for n random variates $X_1, \ldots, X_n$ is defined as the joint density $f_{X_1,\ldots,X_n}(x_1, \ldots, x_n; \boldsymbol{\theta})$, seen as a function of parameter vector $\boldsymbol{\theta}$.

Let $X_1, \ldots, X_n$ be a simple random sample of X. Random variate X has the density distribution $f_X(x, \boldsymbol{\theta})$, then the likelihood function is

$$f_{X_1,\ldots,X_n}(x_1, \ldots, x_n; \boldsymbol{\theta}) = \prod_{i=1}^{n} f_{X_i}(x_i; \boldsymbol{\theta}) = \prod_{i=1}^{n} f_X(x_i; \boldsymbol{\theta}). \tag{1.18}$$

The likelihood function is often denoted by $L(\boldsymbol{\theta}; x_1, \ldots, x_n)$ or $L(\boldsymbol{\theta})$, which in the case of a simple random sample for continuous X is

$$L(\boldsymbol{\theta}; x_1, \ldots, x_n) = L(\boldsymbol{\theta}) = \prod_{i=1}^{n} f_X(x_i; \boldsymbol{\theta}). \tag{1.19}$$

1.3.2 The Exponential Distribution

We look at a very simple example, the exponential distribution. It is a continuous distribution with one parameter only. Let $X \sim \text{Exp}(\lambda)$ with density

$$f(x; \lambda) = \begin{cases} \lambda\, \mathrm{e}^{-\lambda x}, & \text{for } x \geq 0 \\ 0, & \text{for } x < 0. \end{cases} \tag{1.20}$$

The likelihood and log-likelihood functions are given by

$$L(\lambda; x_1, \ldots, x_n) = \prod_{i=1}^{n} \lambda\, e^{-\lambda x_i} \qquad \text{and}$$
$$\ln L(\lambda; x_1, \ldots, x_n) = n \ln \lambda - \lambda \sum_{i=1}^{n} x_i. \tag{1.21}$$

To find the maximum we obtain the derivatives and equate them to 0

$$\frac{\partial \ln L(\lambda)}{\partial \lambda} = \frac{n}{\lambda} - \sum_{i=1}^{n} x_i \stackrel{!}{=} 0 \quad \Rightarrow \quad \hat{\lambda} = \frac{n}{\sum_{i=1}^{n} x_i} = \frac{1}{\bar{x}}. \tag{1.22}$$

The ml-estimator of the unknown parameter λ is

$$\hat{\lambda} = \frac{1}{\bar{X}}. \tag{1.23}$$

To illustrate the basic idea of the ml-estimation using R, we generate a small artificial data set and carry out a grid search using the function `optimize()`.[3]

```
set.seed(123)
n <- 20
x <- rexp(n,rate=5)
# Definition of likelihood functions
likelihood <- function(lambda){
  prod(lambda*exp(-lambda*x))
}
loglikelihood <- function(lambda){
  n*log(lambda)-lambda*sum(x)
}
# plotting the likelihood
lv <- seq(3,10,0.01)
lik <- rep(NA,length(lv))
for(i in 1:length(lv)) lik[i] <- likelihood(lv[i])
no <- which(lik==max(lik))
no

## [1] 317
```

The element of the object `lv` at position 317 is

```
lv[no]

## [1] 6.16
```

[3]Note that `log()` in R computes the natural logarithm.

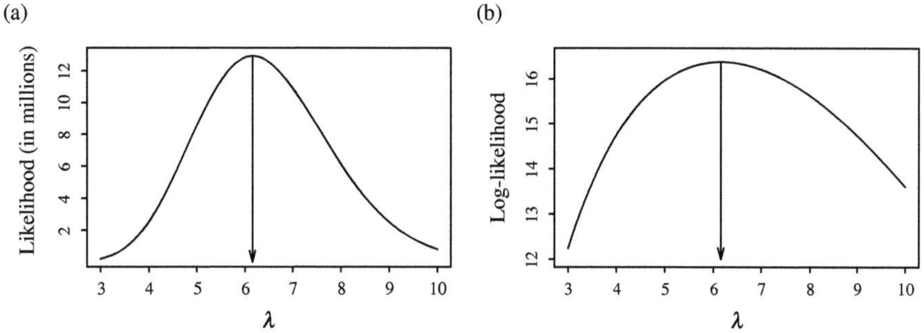

Fig. 1.2 Maximum likelihood estimation. (**a**) Likelihood. (**b**) Log-likelihood

Because the logarithm is a monotonic transformation the likelihood and the log-likelihood have their maximal values at identical parameter values. We can plot the likelihood and the log-likelihood as follows (see Fig. 1.2).

```
# likelihood
plot(lv,lik,type="l",xlab=expression(lambda),
     ylab="Likelihood (in millions)",yaxt='none')
axis(2,at=c(2e+6,4e+6,6e+6,8e+6,10e+6,12e+6),
     labels=c(2,4,6,8,10,12))
arrows(lv[no],lik[no],lv[no],0,length=0.15,
       angle=15)
# log-likelihood
plot(lv,log(lik),type="l",xlab=expression(lambda),
     ylab="Log-likelihood",ylim=c(12,16.5))
arrows(lv[no],log(lik[no]),lv[no],12,length=0.15,
       angle=15)
```

The same estimate $\hat{\lambda} = 6.16$, we obtain using the optimize function:

```
optimize(f=likelihood,interval=c(0,10),maximum=TRUE)

## $maximum
## [1] 6.164
##
## $objective
## [1] 12919465

optimize(f=loglikelihood,interval=c(0,10),maximum=TRUE)

## $maximum
## [1] 6.164
##
## $objective
## [1] 16.37
```

For problems involving a parameter vector, we use the function `optim()` for numerical optimization.[4] For example the standard example is a normal random variate:

```
set.seed(123)
n <- 20
x <- rnorm(n,mean=5,sd=2)
loglikelihood <- function(b){
  -sum(dnorm(x,b[1],b[2],log=TRUE))
}
o <- optim(par=c(1,1),fn=loglikelihood,hessian=T)
o$value

## [1] 41.17

o$par

## [1] 5.283 1.895
```

1.3.3 Properties of ml-Estimators

If regularity conditions are met, ml-estimators have the following properties:

- **Invariance:**
  If $\hat{\boldsymbol{\theta}}$ is the ml-estimator of $\boldsymbol{\theta}$, then $\mathbf{g}(\hat{\boldsymbol{\theta}})$ is the ml-estimator of $\mathbf{g}(\boldsymbol{\theta})$. If the parameter is a continuous differentiable function of θ_0 (true parameter), then this function has its maximum at $\hat{\theta}$.
- **Consistency:**

$$\mathrm{P}(\lim_{n\longrightarrow\infty} \hat{\theta} = \theta_0) = 1.$$

  That is with n going to infinity the probability mass will concentrate on the true parameter value.
- **Asymptotic normality:**

$$\sqrt{n}\left(\hat{\boldsymbol{\theta}}_n - \boldsymbol{\theta}\right) \rightarrow_{\mathrm{d}} \mathbf{U} \sim \mathcal{N}(\mathbf{0}, \mathbf{V}(\boldsymbol{\theta}))$$

$$\hat{\theta} \overset{\mathrm{a}}{\sim} \mathcal{N}\left[\theta_0, \{I(\theta_0)\}^{-1}\right],$$

  where

$$I(\theta_0) = -\,\mathrm{E}_0\left[\partial^2 \ln L/\partial\theta_0\partial\theta_0'\right].$$

[4]Because `optim()` by default minimizes we define a negative likelihood.

- **Asymptotic efficiency:**
  $\mathbf{V}(\boldsymbol{\theta})$ is Cramér–Rao-lower bound. In the class of consistent estimators no other estimator has smaller variance (given n is going to infinity).
- **Feasibility:**
  Even if there is no analytic solution to the ml-problem, one can use numerical optimization methods to obtain the estimate.

We use the following notation:

- $\ln f(y_i|\theta)$ is the value of the log-likelihood function at y_i.
- $g_i = \partial \ln f(y_i|\theta)/\partial\theta$ is the value of the derivative of the log-likelihood function at y_i.
- $H_i = \partial^2 \ln f(y_i|\theta)/\partial\theta\partial\theta'$ is the value of the second order derivative of the log-likelihood function at y_i.

The derivatives of the log-likelihood function have the following properties

- $\mathrm{E}_0\left[g_i(\theta_0)\right] = 0$
  The expected value for derivatives evaluated at y_i is 0 for the true parameter.
- $\mathrm{Var}\left[g_i(\theta_0)\right] = -\mathrm{E}\left[H_i(\theta_0)\right]$

The variance of the derivatives evaluated at y_i equals the negative expected value of the second derivatives evaluated at y_i. The Fisher Information provides information on the quality of the ml-estimator. The second derivative tells about the shape (curvature) of the likelihood. The more the curvature is bend (changing slope of the log-likelihood), the better we can locate the maximum of the log-likelihood and find the estimate $\hat{\theta}$ for θ. The precision of the estimation is formally expressed through the variance of the estimator. The smaller the variance, the more reliable is the obtained estimator. In the maximum the second derivative is negative. Hence, taking the negative second derivative results in a positive value

$$\begin{aligned} I(\theta_0) &= -\mathrm{E}\left[H(\theta_0)\right] = -\mathrm{E}\left[\frac{\partial^2 \ln L(\theta_0)}{\partial\theta_0\partial\theta_0'}\right] \\ &= \mathrm{E}\left[\left(\frac{\partial \ln L(\theta_0)}{\partial\theta_0}\right)\left(\frac{\partial \ln L(\theta_0)}{\partial\theta_0}\right)'\right]. \end{aligned} \tag{1.24}$$

We need the expected value of the second derivatives of the log-likelihood evaluated at y_i under the true parameter. This equals the positive expected value of the first derivatives of the log-likelihood evaluated at y_i under the true parameter. The inverse of the Fisher information matrix is a lower bound (Rao–Cramer) for the variance of the ml-estimator:

$$\{I(\theta_0)\}^{-1} \leq \mathrm{Var}(\hat{\theta}). \tag{1.25}$$

Asymptotically, the variance of the ml-estimator approaches the inverse of the Fisher information matrix

$$\begin{aligned}\mathrm{Avar}(\hat{\theta}) = \{I(\theta_0)\}^{-1} &= \left\{-\mathrm{E}\left[\frac{\partial^2 \ln L(\theta_0)}{\partial\theta_0\partial\theta_0'}\right]\right\}^{-1} \\ &= \left\{\mathrm{E}\left[\left(\frac{\partial \ln L(\theta_0)}{\partial\theta_0}\right)\left(\frac{\partial \ln L(\theta_0)}{\partial\theta_0}\right)'\right]\right\}^{-1}.\end{aligned} \tag{1.26}$$

Note that when estimating we do not know the true parameter θ_0 and therefore we cannot obtain the expected value under the true parameter. Usually, we also cannot obtain the expected value. To circumvent these problems, we use the ml-estimator $\hat{\theta}$ instead of the true parameter θ_0 and we only evaluate at $\hat{\theta}$ instead of taking the expectation for all possible values. We can either approximate the information matrix using the second derivatives

$$\hat{I}(\hat{\theta}) = -\left[\frac{\partial^2 \ln L(\hat{\theta})}{\partial\hat{\theta}\partial\hat{\theta}'}\right], \tag{1.27}$$

or (more often, as second derivatives may be difficult to obtain) using the first derivatives

$$\hat{\hat{I}}(\hat{\theta}) = \left(\frac{\partial \ln L(\hat{\theta})}{\partial\hat{\theta}}\right)\left(\frac{\partial \ln L(\hat{\theta})}{\partial\hat{\theta}}\right)'. \tag{1.28}$$

Because asymptotically the variance approaches the inverse of the Fisher information matrix, we can obtain confidence intervals

$$\mathrm{P}\left(\hat{\theta} - u_{1-\alpha/2}\sqrt{I(\theta)^{-1}} \leq \theta \leq \hat{\theta} + u_{1-\alpha/2}\sqrt{I(\theta)^{-1}}\right) = 1-\alpha. \tag{1.29}$$

Our estimate is based on the evaluated Fisher information matrix

$$\mathrm{P}\left(\hat{\theta} - u_{1-\alpha/2}\sqrt{\hat{\hat{I}}(\hat{\theta})^{-1}} \leq \theta \leq \hat{\theta} + u_{1-\alpha/2}\sqrt{\hat{\hat{I}}(\hat{\theta})^{-1}}\right) = 1-\alpha. \tag{1.30}$$

The asymptotic covariance matrix of the ml-estimator $\hat{\boldsymbol{\theta}}$ is

$$\mathrm{Cov}(\hat{\boldsymbol{\theta}}) = \left[-\mathrm{E}\left(\frac{\partial^2 \ln L(\boldsymbol{\theta}_0)}{\partial\boldsymbol{\theta}_0\partial\boldsymbol{\theta}_0'}\right)\right]^{-1} \tag{1.31}$$

and can consistently be estimated by

$$\widehat{\text{Cov}}(\hat{\boldsymbol{\theta}}) = -\left(\frac{\partial^2 \ln L(\hat{\boldsymbol{\theta}})}{\partial \hat{\boldsymbol{\theta}} \partial \hat{\boldsymbol{\theta}}'}\right)^{-1}. \tag{1.32}$$

Note that the Hessian matrix of the log-likelihood function can usually be estimated using numerical optimization routines.

For our small numerical example, we use the Hessian matrix to construct a confidence interval with $\alpha = 0.05$ for μ:

```
h <- o$hessian
h

##              [,1]        [,2]
## [1,] 5.5665485   0.0008202
## [2,] 0.0008202 11.1433741

hi <- solve(h)
hi

##                [,1]         [,2]
## [1,]  0.17964453 -0.00001322
## [2,] -0.00001322  0.08973943

s <- sqrt(diag(hi))
s

## [1] 0.4238 0.2996

int_l <- o$par[1]+s[1]*qt(0.975,n-2)
int_l

## [1] 6.174

int_h <- o$par[1]-s[1]*qt(0.975,n-2)
int_h

## [1] 4.393
```

Finally, we show the curvature of the log-likelihood by drawing a perspective plot (see Fig. 1.3).

```
m <- 40
mv <- seq(mean(x)-1.5*sd(x)/sqrt(n),
          mean(x)+1.5*sd(x)/sqrt(n),length.out=m)
sv <- seq(0.8*sd(x),1.5*sd(x),length.out=m)
z <- matrix(NA,m,m)
loglikelihood <- function(b){
  -sum(dnorm(x,b[1],b[2],log=TRUE))
}
for(i in 1:m){
  for(j in 1:m){
    z[i,j] <- -loglikelihood(c(mv[i],sv[j]))
    }
```

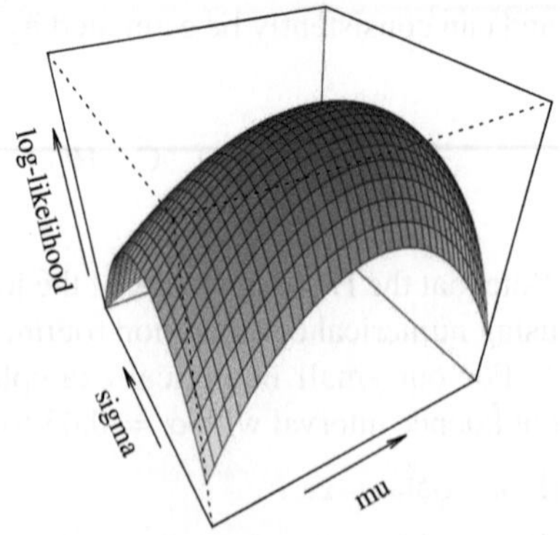

Fig. 1.3 Curvature of the log-likelihood

```
  }
jet.colors <- colorRampPalette(c("grey100", "grey40"))
nbcol <- 100
color <- jet.colors(nbcol)
zfacet <- z[-1, -1] + z[-1, -m] + z[-m, -1] + z[-m, -m]
facetcol <- cut(zfacet, nbcol)
# plot
# "\n" adds one line before the label
persp(mv,sv,z,
      xlab="\n mu",ylab="\n sigma",
      zlab="\n log-likelihood",
      phi=35,theta=-30,col=color[facetcol])
```

1.4 Numerical Optimization

1.4.1 The General Form of Algorithms

We assume that we want to minimize function $H(\theta)$ (e.g., the sum of squared residuals). Our aim is to vary the parameter vector θ in a sensible way to find the minimum. θ_n denotes the parameter vector in the nth-step of the iteration. We update the parameter vector in the next step towards θ_{n+1} according to the formula

$$\theta_{n+1} = \theta_n + \lambda_n \Delta_n, \qquad H(\theta_{n+1}) < H(\theta_n) \tag{1.33}$$

where λ_n is a scalar setting the length of the parameter adjustment. Δ_n denotes the vector of directions into which the parameter vector is updated. We stop the iteration process if a previously specified criterion is reached, e.g.

$$(\theta_{n+1} - \theta_n)' (\theta_{n+1} - \theta_n) < \varepsilon. \tag{1.34}$$

The vector Δ_n is the product of gradient z_n of $H(\theta)$ evaluated at θ_n and the positive definite matrix P_n

$$\Delta_n = -P_n z_n. \tag{1.35}$$

The gradient is given by

$$\left.\frac{\partial H(x,\theta)}{\partial \theta}\right|_{\theta_n} = z_n. \tag{1.36}$$

Inserting into the equation above, we obtain

$$\theta_{n+1} = \theta_n - \lambda_n P_n z_n. \tag{1.37}$$

Many of the well-known algorithms follow this principle idea and differ only in the choice of the matrix P_n.

1.4.2 The Newton–Raphson Algorithm

This algorithm is based on a second order Taylor series approximation of the objective function $H(\theta)$. The function $H(\theta)$ at θ can be approximated through the expansion of the function around θ_n

$$\begin{aligned} H(\theta) \approx H(\theta_n) &+ \left.\frac{\partial H(x,\theta)}{\partial \theta}\right|_{\theta_n} (\theta - \theta_n) \\ &+ \frac{1}{2}(\theta - \theta_n)' \left.\frac{\partial^2 H(x,\theta)}{\partial \theta \partial \theta'}\right|_{\theta_n} (\theta - \theta_n). \end{aligned} \tag{1.38}$$

We obtain the derivative of the approximation of $H(\theta)$ with respect to θ and set the derivative to zero

$$\left.\frac{\partial H(x,\theta)}{\partial \theta}\right|_{\theta_n} + \left.\frac{\partial^2 H(x,\theta)}{\partial \theta \partial \theta'}\right|_{\theta_n} (\theta - \theta_n) = 0. \tag{1.39}$$

Manipulating the expression results in the rule for updating the parameter vector:

$$\theta_{n+1} = \theta_n - \left(\left.\frac{\partial^2 H(x,\theta)}{\partial \theta \partial \theta'}\right|_{\theta_n} \right)^{-1} \left.\frac{\partial H(x,\theta)}{\partial \theta}\right|_{\theta_n} \tag{1.40}$$

where the weighting matrix P_n is the inverse of the matrix of second derivatives (Hessian matrix).

1.4.3 Two Simple Examples

We illustrate the algorithm using a simple nonlinear regression function

$$y = f(x) + \varepsilon = x^{\theta} + \varepsilon. \tag{1.41}$$

The function $H(\theta)$ is the sum of squared residuals

$$H(\theta) = \sum_{i=1}^{n} [y_i - f(x_i)]^2 = \sum_{i=1}^{n} [y_i - x_i^{\theta}]^2. \tag{1.42}$$

Derivatives of $H(\theta)$ are given by

$$\frac{\partial H(\theta)}{\partial \theta} = 2 \sum_{i=1}^{n} [y_i - x_i^{\theta}] \left(-(\ln x_i) x_i^{\theta}\right) \tag{1.43}$$

and

$$\begin{aligned} \frac{\partial^2 H(\theta)}{\partial \theta^2} = 2 \sum_{i=1}^{n} \Big\{ &\left(-(\ln x_i) x_i^{\theta}\right) \left(-(\ln x_i) x_i^{\theta}\right) \\ &+ \left(y_i - x_i^{\theta}\right) \left(-(\ln x_i)^2 x_i^{\theta}\right) \Big\}. \end{aligned} \tag{1.44}$$

A very simple numerical example with two observations is

$$x_1 = 1, x_2 = 2, y_1 = 1, y_2 = 2$$

Starting with $\theta_0 = 0.5$, we find

$$H(\theta)|_{\theta_0} = \left[1 - 1^{0.5}\right]^2 + \left[2 - 2^{0.5}\right]^2 = 0.34315. \tag{1.45}$$

For the derivatives, we find

$$\begin{aligned} \frac{\partial H(\theta)}{\partial \theta} &= 2 \sum_{i=1}^{n} [y_i - x_i^{\theta}] \left(-(\ln x_i) x_i^{\theta}\right) \\ &= 2 \cdot \left(\left(1 - 1^{0.5}\right) \left(-(\ln 1) 1^{0.5}\right) + \left(2 - 2^{0.5}\right) \left(-(\ln 2) 2^{0.5}\right)\right) \\ &= -1.1485 \end{aligned} \tag{1.46}$$

and

$$\frac{\partial^2 H(\theta)}{\partial \theta^2}|_{\theta_0} = 2 \cdot \{ \left(-\ln 1 \cdot 1^{0.5}\right)^2 + \left(1 - 1^{0.5}\right)(-\ln 1)^2 \cdot 1^{0.5} \\ + \left((-\ln 2) \cdot 2^{0.5}\right)^2 + \left(2 - 2^{0.5}\right)(-\ln 2)^2 \cdot 2^{0.5}\} \\ = 2.7179. \tag{1.47}$$

For θ_1 we obtain

$$\theta_1 = 0.5 - \frac{1}{2.7179} \cdot (-1.1485) = 0.92257. \tag{1.48}$$

The next iterations will result in $\theta_2 = 0.9979581$, $\theta_3 = 0.9999986$ and finally $\theta_4 = 1$ for which we find

$$H(\theta)|_{\theta_4} = \left[1 - 1^1\right]^2 + \left[2 - 2^1\right]^2 = 0. \tag{1.49}$$

We solve the numerical optimization problem with R:

```
x <- 1:2
y <- 1:2
b0 <- 0.5
H <- function(b){sum((y-x^b)^2)}
H1 <- function(b){2*sum(  (y-x^b)*(-log(x)*x^b)  )}
H2 <- function(b){2*sum((-log(x)*x^b)^2
                          + (y-x^b)*(-log(x))^2*x^b)}
H(b0)

## [1] 0.34315

H1(b0)

## [1] -1.1484

H2(b0)

## [1] 2.7179

b1 <- b0-H2(b0)^(-1)*H1(b0)
b1

## [1] 0.92256

b2 <- b1-H2(b1)^(-1)*H1(b1)
b2

## [1] 0.99796

b3 <- b2-H2(b2)^(-1)*H1(b2)
b3

## [1] 1

H(b3)

## [1] 4.0091e-12
```

R provides several algorithms for numerical optimization implemented in the function `optim()`. We use the BFGS-algorithm named after Broyden, Fletcher, Goldfarb, and Shanno which is a quasi-Newton method and provide the first derivative[5]:

```
o <- optim(c(0.5),H,H1,method="BFGS")
o$par

## [1] 1

o$value

## [1] 7.0997e-30
```

We illustrate the algorithm using a second simple nonlinear regression example

$$y = f(x) + \varepsilon = \theta_1 + x^{\theta_2} + \varepsilon. \tag{1.50}$$

The function $H(\theta)$ in this case is

$$H(\theta) = \sum_{i=1}^{n} [y_i - f(x_i)]^2 = \sum_{i=1}^{n} \left[y_i - \theta_1 - x_i^{\theta_2}\right]^2. \tag{1.51}$$

Derivatives of $H(\theta)$ are given by

$$\frac{\partial H(\theta)}{\partial \theta_1} = \frac{\partial \sum_{i=1}^{n} \left[y_i - \theta_1 - x_i^{\theta_2}\right]^2}{\partial \theta_1} = 2\sum_{i=1}^{n} \left(\theta_1 - y_i + x_i^{\theta_2}\right), \tag{1.52a}$$

$$\frac{\partial H(\theta)}{\partial \theta_2} = \frac{\partial \sum_{i=1}^{n} \left[y_i - \theta_1 - x_i^{\theta_2}\right]^2}{\partial \theta_2} = 2\sum_{i=1}^{n} x_i^{\theta_2} (\ln x_i) \left(\theta_1 - y_i + x_i^{\theta_2}\right), \tag{1.52b}$$

$$\frac{\partial^2 H(\theta)}{\partial \theta_1^2} = 2n, \tag{1.52c}$$

$$\frac{\partial^2 H(\theta)}{\partial \theta_2^2} = 2\sum_{i=1}^{n} x_i^{\theta_2} (\ln^2 x_i) \left(\theta_1 - y_i + 2x_i^{\theta_2}\right) \quad \text{and} \tag{1.52d}$$

$$\frac{\partial^2 H(\theta)}{\partial \theta_1 \partial \theta_2} = 2\sum_{i=1}^{n} x_i^{\theta_2} \ln x_i. \tag{1.52e}$$

[5]The Newton–Raphson algorithm is available in the maxLik package.

A very simple numerical example with three observations is

$$x_1 = 1, \quad x_2 = 2, \quad x_3 = 3, \quad y_1 = 2, \quad y_2 = 2.414, \quad y_3 = 2.732.$$

Starting with $\theta_0 = (2, 0.2)'$ for the function $H(\theta)$ we find

$$\begin{aligned} H(\theta) &= \sum_{i=1}^{n} \left[y_i - \theta_1 - x_i^{\theta_2} \right]^2 = \left(2 - 2 - 1^{0.2}\right)^2 + \left(2.414 - 2 - 2^{0.2}\right)^2 \\ &\quad + \left(2.732 - 2 - 3^{0.2}\right)^2 = 1.8037. \end{aligned} \tag{1.53}$$

For the derivatives we find

$$\begin{aligned} \frac{\partial H(\theta)}{\partial \theta_1} &= -2 \sum_{i=1}^{n} \left[y_i - \theta_1 - x_i^{\theta_2} \right] \\ &= -2\{\left(2 - 2 - 1^{0.2}\right) + \left(2.414 - 2 - 2^{0.2}\right) \\ &\quad + \left(2.732 - 2 - 3^{0.2}\right)\} = 4.4969, \end{aligned} \tag{1.54a}$$

$$\begin{aligned} \frac{\partial H(\theta)}{\partial \theta_2} &= -2 \sum_{i=1}^{n} x_i^{\theta_2} (\ln x_i) \left[y_i - \theta_1 - x_i^{\theta_2} \right] \\ &= -2\{1^{0.2} \cdot \ln 1 \cdot \left(2 - 2 - 1^{0.2}\right) \\ &\quad + 2^{0.2} \cdot \ln 2 \cdot \left(2.414 - 2 - 2^{0.2}\right) \\ &\quad + 3^{0.2} \cdot \ln 3 \cdot \left(2.732 - 2 - 3^{0.2}\right)\} \\ &= -2(0 + 0.79622 \cdot (-0.73470) \\ &\quad + 1.3686 \cdot (-0.51373)) \\ &= 2.5761, \end{aligned} \tag{1.54b}$$

$$\frac{\partial^2 H(\theta)}{\partial \theta_1^2} = 2n = 6, \tag{1.54c}$$

$$\begin{aligned} \frac{\partial^2 H(\theta)}{\partial \theta_2^2} &= -2 \sum_{i=1}^{n} x_i^{\theta_2} (\ln^2 x_i) \left(y_i - \theta_1 - 2x_i^{\theta_2} \right) \\ &= -2\{1^{0.2} \cdot \ln^2 1 \cdot \left(2 - 2 - 2 \cdot 1^{0.2}\right) \\ &\quad + 2^{0.2} \cdot \ln^2 2 \cdot \left(2.414 - 2 - 2 \cdot 2^{0.2}\right) \\ &\quad + 3^{0.2} \cdot \ln^2 3 \cdot \left(2.732 - 2 - 2 \cdot 3^{0.2}\right)\} \\ &= -2 \cdot (0 + 0.55190 \cdot (-1.8834) \\ &\quad + 1.5035 \cdot (-1.7595)) = 7.3697 \quad \text{and} \end{aligned} \tag{1.54d}$$

$$
\begin{aligned}
\frac{\partial^2 H(\theta)}{\partial \theta_1 \partial \theta_2} &= 2\sum_{i=1}^{n} x_i^{\theta_2} \ln x_i \\
&= 2 \cdot \left(1^{0.2} \cdot \ln 1 + 2^{0.2} \cdot \ln 2 + 3^{0.2} \cdot \ln 3\right) \\
&= 4.3296.
\end{aligned} \tag{1.54e}
$$

For θ_1, we obtain

$$
\begin{aligned}
\theta_{n+1} &= \theta_n - \left(\left. \frac{\partial^2 H(x,\theta)}{\partial \theta \partial \theta'} \right|_{\theta_n} \right)^{-1} \left. \frac{\partial H(x,\theta)}{\partial \theta} \right|_{\theta_n} \\
\theta_1 &= \begin{bmatrix} 2 \\ 0.2 \end{bmatrix} - \begin{bmatrix} 6 & 4.3296 \\ 4.3296 & 7.3697 \end{bmatrix}^{-1} \begin{bmatrix} 4.4969 \\ 2.5761 \end{bmatrix} \\
&= \begin{bmatrix} 2 \\ 0.2 \end{bmatrix} - \begin{bmatrix} 0.28932 & -0.16997 \\ -0.16997 & 0.23555 \end{bmatrix} \begin{bmatrix} 4.4969 \\ 2.5761 \end{bmatrix} = \begin{bmatrix} 1.1368 \\ 0.3575 \end{bmatrix}.
\end{aligned} \tag{1.55}
$$

The next iterations will result in $\theta_2 = (0.97571, 0.5391)'$, $\theta_3 = (0.9977, 0.5031)'$ and finally $\theta_4 = (1, 0.5)'$ for which we find

$$
\begin{aligned}
H(\theta)\,|_{\theta_5} = & \left(2 - 1 - 1^{0.5}\right)^2 + \left(2.414 - 1 - 2^{0.5}\right)^2 \\
& + \left(2.732 - 1 - 3^{0.5}\right) = 0.
\end{aligned} \tag{1.56}
$$

```
x <- 1:3
b <- c(1,0.5)
n <- length(x)
f <- function(x) b[1]+x^b[2]
y <- f(x)
H <- function(b){sum((y-b[1]-x^b[2])^2)}
H1 <- function(b){
  matrix(c( 2*sum( b[1]-y+x^b[2] ),
            2*sum(x^b[2]*log(x)*(b[1]-y+x^b[2]))))}
H2 <- function(b){
  matrix(c( 2*n,
            2*sum(x^b[2]*log(x)),
            2*sum(x^b[2]*log(x)),
            2*sum(x^b[2]*log(x)^2*(b[1]-y+2*x^b[2]))),
            2,2)}
b0 <- c(2,0.2)
H(b0)%%

## [1] 1.8033

H1(b0)
```

```
##          [,1]
## [1,] 4.4963
## [2,] 2.5756
```

```
H2(b0)
```

```
##          [,1]   [,2]
## [1,] 6.0000 4.3296
## [2,] 4.3296 7.3693
```

```
b1 <- matrix(b0)-solve(H2(b0))%*%H1(b0)
b1
```

```
##           [,1]
## [1,] 1.13691
## [2,] 0.35757
```

```
b2 <- matrix(b1)-solve(H2(b1))%*%H1(b1)
b2
```

```
##           [,1]
## [1,] 0.97572
## [2,] 0.53913
```

```
b3 <- matrix(b2)-solve(H2(b2))%*%H1(b2)
b3
```

```
##           [,1]
## [1,] 0.99765
## [2,] 0.50308
```

```
b4 <- matrix(b3)-solve(H2(b3))%*%H1(b3)
b4
```

```
##           [,1]
## [1,] 0.99999
## [2,] 0.50002
```

```
H(b4)
```

```
## [1] 0.00000000066655
```

Using `optim()` and providing the first derivative for the BFGS-algorithm, we obtain:

```
o <- optim(b0,H,H1,method="BFGS")
o$par
```

```
## [1] 1.0 0.5
```

```
o$value
```

```
## [1] 5.3571e-18
```

1.5 Recommended Readings

The R software for statistical computing is now used in many faculties and introductory courses are given at most universities. The R homepage (www.r-project.org) provides manuals and an exhaustive list of R-books.

- R Core Team (2013) R: A language and environment for statistical computing. R Foundation for Statistical Computing, Vienna, Austria, URL http://www.R-project.org/

An excellent book which provides concise overviews to many different areas of statistics is

- Venables BD William N; Ripley (2002) Modern applied statistics with S, 4th edn. Springer, New York

Some familiarity with regression analysis is assumed throughout the book, but mainly so in the chapters on production functions (with cross sectional and with panel data) and stochastic frontier analysis (again with cross sectional and with panel data). A very recommendable introduction to regression analysis with R is

- Fox J, Weisberg S (2011) An R companion to applied regression. Sage Publications, Thousand Oaks

Reference textbooks on econometrics are:

- Greene WH (2003) Econometric analysis, 5th edn. Pearson Education, New Jersey
- Johnston J, DiNardo J (1997) Econometric methods, 4th edn. R. R. Donnelley & Sons

1.6 Exercises

1. Define a vector x containing the values $\{1, 3, 5, 12\}$.
   a. Sort x using `sort()` and `order()` commands.
   b. Calculate $\sum_i x_i$.
   c. Let object n contain the length of x.
   d. Calculate the variance of x and compare your result with the result obtained from `var(x)`.
   e. Calculate the median of x and compare your results with the results obtained from the following commands:

```
quantile(x,type=1,0.5)
quantile(x,type=7,0.5)
median(x)
```

2. Probability distributions
   a. Consider the exponential distribution with rate $\lambda = 0.2$.
      i. Calculate the value of the probability distribution for $x = 2.5$.
      ii. Calculate the value of the density distribution for $x = 0.8$.

iii. Calculate the median.
iv. Generate a realization with $n = 10$ from the exponential distribution (rate $\lambda = 0.2$).

b. Consider the normally distributed random variate $X \sim \mathcal{N}(\mu = 2, \sigma^2 = 5)$ and calculate

$$P(X < 3), \quad P(X > 0.5), \quad \text{and} \quad P(-2 < X < 3).$$

c. Consider the Poisson distributed random variate $X \sim \text{Pois}(\lambda = 4)$ and calculate

$$P(X = 3), \quad P(X \leq 2), \quad \text{and} \quad P(1 < X \leq 5).$$

d. Generate a realization with $n = 100$ from the normal distribution ($X \sim \mathcal{N}(\mu = 2, \sigma^2 = 5)$) and display the distribution using `hist()`.

3. Define the following matrix using the command

```
x <- matrix(1:16,nrow=4,ncol=4,byrow=F)
```

$$x = \begin{pmatrix} 1 & 5 & 9 & 13 \\ 2 & 6 & 10 & 14 \\ 3 & 7 & 11 & 15 \\ 4 & 8 & 12 & 16 \end{pmatrix}.$$

a. Select the vector containing the second column of x.
b. Select the vector containing the third row of x.
c. Select the 2×2-matrix containing the elements $x_{23}, x_{24}, x_{33}, x_{34}$.

4. Consider the following simple numerical example

$$y = \begin{bmatrix} 3 \\ 1 \\ 8 \\ 3 \\ 5 \end{bmatrix} \quad X = \begin{bmatrix} 1 & 3 & 5 \\ 1 & 1 & 4 \\ 1 & 5 & 6 \\ 1 & 2 & 4 \\ 1 & 4 & 6 \end{bmatrix}.$$

a. Obtain the parameter vector $\hat{\beta}$ of the linear model

$$y_i = \beta_0 + \beta_1 x_{1i} + \beta_2 x_{2i} + u_i.$$

b. Test the following hypotheses
i. $H_0 : \beta_1 = 0$,
ii. $H_0 : \beta_1 = 1$ and
iii. $H_0 : \beta_1 = 2; \beta_2 = -1$.

5. Consider the following production function

$$y = x_1^{\beta_1} x_2^{\beta_2} + u \quad \text{with} \quad u \sim \mathcal{N}(0, \sigma^2).$$

a. Derive the likelihood.
b. Derive the log-likelihood.
c. Derive the score function. Remember

$$\frac{\partial x^\beta}{\partial \beta} = \frac{\partial \exp(\log x^\beta)}{\partial \beta} = \frac{\partial \exp(\beta \log x)}{\partial \beta} = x^\beta \log x.$$

d. Derive the Hessian matrix.
e. Generate realizations of the stochastic production function using the following setting (`set.seed(123)`)

$$\beta_1 = 0.6, \beta_2 = 0.4, n = 50$$

$$x_1, x_2 \sim \mathcal{U}(0.1, 1), \quad \text{with} \quad u \sim \mathcal{N}(0, 0.05^2)$$

and estimate the parameters using the Newton–Raphson-algorithm.
f. Compare your results with the results obtained from `optim()`.

References

Fox J, Weisberg S (2011) An R companion to applied regression. Sage Publications, Thousand Oaks

Greene WH (2003) Econometric analysis, 5th edn. Pearson Education, New Jersey

Johnston J, DiNardo J (1997) Econometric methods, 4th edn. R. R. Donnelley & Sons

R Core Team (2013) R: A language and environment for statistical computing. R Foundation for Statistical Computing, Vienna, Austria, URL http://www.R-project.org/

Venables BD William N; Ripley (2002) Modern applied statistics with S, 4th edn. Springer, New York

Linear Production Model

2

Abstract

Linear production models allow a concise representation of the production process in an economy. The interdependence between production units is its main characteristic as each production unit uses the output of other producing units as inputs. Empirical content is provided by analyzing highly aggregated input–output tables for Germany.

2.1 The Theoretical Linear Production Model

In this section, we introduce a simple linear production model to develop an idea of the production process of an economy. The model is mainly based on the works of Schwartz (1961) and Sraffa (1963). Input–output tables allow to bring some empirical content to the linear production models. We analyze some input–output tables for Germany for 1991 and 2007.

2.1.1 Industries and Goods

We assume that in our economy, there are n production units (firms or sectors) $\omega_1, \ldots, \omega_n$. A strong simplification is the important assumption that the production unit ω_i produces exactly one kind of good c_i and that good c_i is only produced by unit ω_i. Therefore, corresponding to the n production units, we have $c_1, \ldots, c_n$ different goods. x_i is the quantity of good c_i that is produced by production unit ω_i. Hence, the quantities produced in the economy $x_1, \ldots, x_n$ are given by the following

A. Behr, *Production and Efficiency Analysis with R*,
DOI 10.1007/978-3-319-20502-1_2

vector $\mathbf{x}$[1]

$$\mathbf{x} := (x_1, \ldots, x_n)'. \tag{2.1}$$

As the quantity produced depends on the length of the production period, we assume in the following that the length of the production period is one year. To produce output of good c_j, the producing unit ω_j needs other goods as inputs. The input matrix $\mathbf{A}$ contains the elements a_{ij} stating how much input of kind i the production unit j used

$$\mathbf{A} = \begin{bmatrix} a_{11} & \cdots & a_{1n} \\ \vdots & \ddots & \vdots \\ a_{n1} & \cdots & a_{nn} \end{bmatrix}. \tag{2.2}$$

π_{ij} denotes the quantity of input of good c_i that production unit ω_j needs to produce one unit of c_j. For the input coefficients it holds that $\pi_{ij} \geq 0$. If $\pi_{ij} = 0$, no input c_i is necessary to produce good c_j. Note that usually $\pi_{ii} > 0$, which means that producing units use some of their own output as inputs.[2] Therefore, the production process in production unit ω_j is characterized by its input coefficients. If production unit ω_j wants to produce the quantity x_j of good c_j, it has to use the quantities $\pi_{1j}x_j, \ldots, \pi_{nj}x_j$ as inputs.

x_i is the quantity of good c_i that is produced in the production period. Thereof, the quantities $\pi_{i1}x_1, \ldots, \pi_{in}x_n$ are used as inputs in the n producing units. Hence, only a fraction of the produced quantity of x_i is available for final demand y_i. This can be formulated using matrix notation. We combine all input coefficients π_{ij} in the production matrix $\boldsymbol{\Pi}$

$$\boldsymbol{\Pi} := \begin{bmatrix} \pi_{11} & \cdots & \pi_{1n} \\ \vdots & & \vdots \\ \pi_{n1} & \cdots & \pi_{nn} \end{bmatrix}. \tag{2.3}$$

If we multiply the production matrix with the vector of total output $\mathbf{x}$, we obtain a vector containing the quantities of the n goods that have to be used as inputs in the production process

$$\boldsymbol{\Pi}\,\mathbf{x} = \begin{bmatrix} \sum_{j=1}^{n} \pi_{1j}\, x_j \\ \vdots \\ \sum_{j=1}^{n} \pi_{nj}\, x_j \end{bmatrix}. \tag{2.4}$$

[1] $\mathbf{x}$ is a column vector. The prime $'$ indicates the transposition of the vector.

[2] For example wheat used as input in agriculture.

If we subtract the amounts used as inputs from the quantities produced, we obtain the vector $\mathbf{y}$ containing the quantities that are available for final demand

$$\mathbf{y} := \mathbf{x} - \boldsymbol{\Pi}\,\mathbf{x} = (\mathbf{I} - \boldsymbol{\Pi})\,\mathbf{x}. \tag{2.5}$$

The symbol $\mathbf{I}$ denotes the unit matrix of dimension (n, n).

Equation (2.5) displays the connection between gross production $\mathbf{x}$ and net production $\mathbf{y}$ (final demand). Inverting the matrix $(\mathbf{I} - \boldsymbol{\Pi})$ allows to calculate the gross production that has to be produced if a vector of final demand $\mathbf{y}$ shall be available for consumption and investment

$$\mathbf{x} = (\mathbf{I} - \boldsymbol{\Pi})^{-1}\,\mathbf{y}. \tag{2.6}$$

The matrix $(\mathbf{I} - \boldsymbol{\Pi})^{-1}$ is known as the Leontief-inverse.

$(\mathbf{I} - \boldsymbol{\Pi})$ has a non-negative inverse if the dominant eigenvalue of $\boldsymbol{\Pi}$ is smaller 1. This can be stated as follows:

> If the production matrix is non-negative and there is a vector of net production which is non-negative having at least one positive element, then the production matrix will have a dominant eigenvalue smaller 1.

Usually, these requirements are met as otherwise all the units would use the entire output as input without delivering anything to other economic units.[3]

2.1.2 Direct and Indirect Inputs

π_{ij} is the quantity of good i that must be used as input to produce one unit of good j. But the goods used as inputs have to be produced by means of inputs as well. Obviously, we can follow the production process backwards infinitely: inputs, inputs of inputs, inputs of inputs of inputs, …In the following, we examine how we can find the sum of all the direct and indirect amounts of inputs.

$\pi_{ij}^{(1)} := \pi_{ij}$ denotes the direct inputs
$\pi_{ij}^{(2)} := \sum_{l=1}^{n} \pi_{il}^{(1)} \pi_{lj}$ denotes the indirect inputs (second stage)
$\pi_{ij}^{(k)} := \sum_{l=1}^{n} \pi_{il}^{(k-1)} \pi_{lj}$ denotes the indirect inputs at the k-th stage of production

Summing up, we find

$$\bar{\pi}_{ij} := \sum_{k=1}^{\infty} \pi_{ij}^{(k)} \tag{2.7}$$

[3]For a formal treatment see Behr and Rohwer (2013), p. 320–323.

as the total amount of direct and indirect inputs. Combining the coefficients $\bar{\pi}_{ij}$ in the matrix $\bar{\boldsymbol{\Pi}}$, we find

$$\bar{\boldsymbol{\Pi}} = \sum_{k=1}^{\infty} \boldsymbol{\Pi}^k. \tag{2.8}$$

Using the following expression for series

$$(\mathbf{I} - \boldsymbol{\Pi})^{-1} = \sum_{k=0}^{\infty} \boldsymbol{\Pi}^k \tag{2.9}$$

we find

$$(\mathbf{I} - \boldsymbol{\Pi})^{-1} = \sum_{k=0}^{\infty} \boldsymbol{\Pi}^k = \mathbf{I} + \sum_{k=1}^{\infty} \boldsymbol{\Pi}^k = \mathbf{I} + \bar{\boldsymbol{\Pi}} \tag{2.10}$$

and

$$\bar{\boldsymbol{\Pi}} = (\mathbf{I} - \boldsymbol{\Pi})^{-1} - \mathbf{I}. \tag{2.11}$$

The matrix $(\mathbf{I} - \boldsymbol{\Pi})^{-1}$ is the Leontief-inverse already introduced above. If we subtract the identity matrix from the Leontief-inverse, we obtain the matrix containing the coefficients indicating the total amount of inputs (direct and indirect) necessary for the production of one unit of output.

2.1.3 The Employment Model

The vector $\mathbf{b} := (b_1, \ldots, b_n)'$ contains the amount of labor (e.g., the number of employees or the number of hours worked) for the n units of production. b_j is the amount of labor in production unit j. Employment coefficients w_j indicate the quantity of labor b_j per unit of output x_j and are given by

$$w_j = \frac{b_j}{x_j}. \tag{2.12}$$

Total employment in the economy b is obtained through summing up employment over the n production units[4]:

$$b = \mathbf{w}'\mathbf{x} = \sum_{j=1}^{n} w_j x_j = \sum_{j=1}^{n} b_j. \tag{2.13}$$

[4] b (scalar) denotes the total amount of labor, $\mathbf{b}$ (vector) denotes the amount of labor in the n production units.

Because of

$$\mathbf{x} = (\mathbf{I} - \boldsymbol{\Pi})^{-1}\mathbf{y} \tag{2.14}$$

we also find

$$b = \mathbf{w}'\mathbf{x} = \mathbf{w}'(\mathbf{I} - \boldsymbol{\Pi})^{-1}\mathbf{y}. \tag{2.15}$$

Total employment is determined through three different factors: $\mathbf{w}$ containing the labor coefficients w_j, the Leontief-inverse $(\mathbf{I} - \boldsymbol{\Pi})^{-1}$, describing the structure of production and the vector of final demand $\mathbf{y}$ containing the quantities of demand for kinds of goods y_j. The coefficients c_{ij} of the Leontief-inverse indicate the quantity of good c_i necessary (direct and indirect) in production unit j to produce one unit of good c_j.

If we build a diagonal matrix diag($\mathbf{w}$) with elements w_j and postmultiply with the vector of outputs $\mathbf{x}$, we obtain a vector of employment by producing units

$$\mathbf{b}' = \mathbf{w}' \operatorname{diag}[(\mathbf{I} - \boldsymbol{\Pi})^{-1}\mathbf{y}] = \mathbf{w}' \operatorname{diag}[\mathbf{x}]. \tag{2.16}$$

Premultiplying the input matrix $\mathbf{A}$ with the diagonalized vector of employment coefficients diag($\mathbf{w}$), we find the input-employment matrix $\mathbf{B}_a$

$$\begin{aligned}\mathbf{B}_a &= \begin{bmatrix} w_1 & \cdots & 0 \\ \vdots & \ddots & \vdots \\ 0 & \cdots & w_n \end{bmatrix} \begin{bmatrix} a_{11} & \cdots & a_{1n} \\ \vdots & \ddots & \vdots \\ a_{n1} & \cdots & a_{nn} \end{bmatrix} \\ &= \begin{bmatrix} w_1 a_{11} & \cdots & w_1 a_{1n} \\ \vdots & \ddots & \vdots \\ w_n a_{n1} & \cdots & w_n a_{nn} \end{bmatrix} = \begin{bmatrix} \frac{b_1}{x_1} a_{11} & \cdots & \frac{b_1}{x_1} a_{1n} \\ \vdots & \ddots & \vdots \\ \frac{b_n}{x_n} a_{n1} & \cdots & \frac{b_n}{x_n} a_{nn} \end{bmatrix}.\end{aligned} \tag{2.17}$$

The elements $w_i a_{ij}$ display the quantity of labor that is necessary to produce input i used in production unit j.

Premultiplying the Leontief-inverse $(\mathbf{I} - \boldsymbol{\Pi})^{-1}$ with the diagonalized vector of employment coefficients diag($\mathbf{w}$), we obtain the employment inverse $\mathbf{H}$

$$\begin{aligned}\mathbf{H} &= \begin{bmatrix} w_1 & \cdots & 0 \\ \vdots & \ddots & \vdots \\ 0 & \cdots & w_n \end{bmatrix} \begin{bmatrix} c_{11} & \cdots & c_{1n} \\ \vdots & \ddots & \vdots \\ c_{n1} & \cdots & c_{nn} \end{bmatrix} \\ &= \begin{bmatrix} w_1 c_{11} & \cdots & w_1 c_{1n} \\ \vdots & \ddots & \vdots \\ w_n c_{n1} & \cdots & w_n c_{nn} \end{bmatrix}.\end{aligned} \tag{2.18}$$

The elements $w_i c_{ij}$ show the direct and indirect amount of labor that is necessary in production unit i to produce the input of good i that is used in production unit j per unit output. If we sum the elements of $\mathbf{H}$ columnwise or premultiply the Leontief-inverse by the row vector of employment coefficients, we obtain employment multipliers for individual goods i. These multipliers $\sum_i^n w_i c_{ij}$ show the total amount of labor that is necessary to produce one unit output of good j and are given by

$$\begin{aligned}
&\begin{bmatrix} w_1, \ldots, w_n \end{bmatrix} \begin{bmatrix} c_{11} & \cdots & c_{1n} \\ \vdots & \ddots & \vdots \\ c_{n1} & \cdots & c_{nn} \end{bmatrix} \\
&= \begin{bmatrix} w_1 c_{11} + w_2 c_{21} + \ldots + w_n c_{n1}, \ldots, w_1 c_{1n} + w_2 c_{2n} + \ldots + w_n c_{nn} \end{bmatrix} \\
&= \begin{bmatrix} \sum_i^n w_i c_{i1}, \ldots, \sum_i^n w_i c_{in} \end{bmatrix}.
\end{aligned} \tag{2.19}$$

2.1.4 A Numerical Example with R

To illustrate the theoretical model, we work through a simple fictitious numerical example. Our economy is supposed to have four production units:

ω_1 produces wood,
ω_2 produces steel,
ω_3 produces cars,
ω_4 produces corn.

The production units produce the following quantities: $x_1 = 100, x_2 = 50, x_3 = 10,$ and $x_4 = 80$. The output vector therefore is

```
x <- matrix(c(100, 50, 10, 80));t(x)

##      [,1] [,2] [,3] [,4]
## [1,]  100   50   10   80
```

Inputs are given in the following matrix:

```
A <- rbind(c(0, 40, 20, 0),
           c(0, 0, 30, 10),
           c(2, 0, 0, 2),
           c(0, 0, 0, 0))

A

##      [,1] [,2] [,3] [,4]
## [1,]    0   40   20    0
## [2,]    0    0   30   10
## [3,]    2    0    0    2
## [4,]    0    0    0    0
```

This reads as follows: ω_1 uses 2 cars to produce wood, ω_2 uses 40 cords[5] of wood in its steel production, and so on.

The input matrix can be transformed to the production matrix $\boldsymbol{\Pi}$ through dividing each column by the output of the respective production unit:

$$\pi_{ij} = \frac{a_{ij}}{x_j}. \tag{2.20}$$

The coefficients π_{ij} state the amount of input i that is in average used for one unit of output j. In our example we find for $\boldsymbol{\Pi}$

```
P <- t(t(A)/as.vector(x));P

##      [,1] [,2] [,3]  [,4]
## [1,] 0.00  0.8    2 0.000
## [2,] 0.00  0.0    3 0.125
## [3,] 0.02  0.0    0 0.025
## [4,] 0.00  0.0    0 0.000
```

and matrix $(\mathbf{I} - \boldsymbol{\Pi})$

```
n <- nrow(P);n

## [1] 4

IP <- diag(n)-P;IP

##       [,1] [,2] [,3]   [,4]
## [1,]  1.00 -0.8   -2  0.000
## [2,]  0.00  1.0   -3 -0.125
## [3,] -0.02  0.0    1 -0.025
## [4,]  0.00  0.0    0  1.000
```

When multiplying the output vector $\mathbf{x}$ from the right to this matrix, we obtain the vector of net production

$$(\mathbf{I} - \boldsymbol{\Pi})\,\mathbf{x} = \mathbf{y}. \tag{2.21}$$

```
y <- IP%*%x;t(y)

##      [,1] [,2] [,3] [,4]
## [1,]   40   10    6   80
```

This is the amount of the output that is not needed as input in the production process and available for final demand.

[5]A standard cord is considered a 4'x4'x8' stack of wood including bark and air space. ' denotes a foot which corresponds to 30,48 cm.

In our numerical example, the requirements for the existence of the inverse are surely met as we see that all units have a positive net production.[6] Hence, we can invert $(\mathbf{I}-\boldsymbol{\Pi})$ and obtain the following non-negative inverse $(\mathbf{I}-\boldsymbol{\Pi})^{-1}$:

```
IPi <- solve(IP);IPi

##        [,1]   [,2] [,3]   [,4]
## [1,] 1.0965 0.8772 4.82 0.2303
## [2,] 0.0658 1.0526 3.29 0.2138
## [3,] 0.0219 0.0175 1.10 0.0296
## [4,] 0.0000 0.0000 0.00 1.0000
```

Subtracting the identity matrix from this inverse, we obtain a matrix containing in each column j the quantities of goods that unit j has to use as inputs to produce one quantity for final demand $(\mathbf{I}-\boldsymbol{\Pi})^{-1}-\mathbf{I}=\bar{\boldsymbol{\Pi}}$:

```
Pbar <- IPi-diag(n);Pbar

##        [,1]   [,2]   [,3]   [,4]
## [1,] 0.0965 0.8772 4.8246 0.2303
## [2,] 0.0658 0.0526 3.2895 0.2138
## [3,] 0.0219 0.0175 0.0965 0.0296
## [4,] 0.0000 0.0000 0.0000 0.0000
```

Let us look at the production unit ω_2 which produces steel. To produce one quantity of steel, the unit uses 0.8 quantities wood as direct input. The total input of wood, comprising direct and indirect inputs, is $\bar{\pi}_{12}=0.88$. Referring to the 50 units of steel, we find that direct input of wood is 40 but direct and indirect input of wood amounts to 44 units of wood.

Assuming a vector with sectoral employment figures

$$\mathbf{b}=(200,200,500,800)'$$

we can calculate sectoral labor coefficients

```
b <- matrix(c(200,200,500,800))
w <- b/x;t(w)

##      [,1] [,2] [,3] [,4]
## [1,]    2    4   50   10
```

Putting these labor coefficients on the diagonal, we can transform the input matrix $\mathbf{A}$ towards the labor input matrix $\mathbf{B}_a$ through premultiplying by the diagonal matrix:

```
Ba <- diag(as.vector(w))%*%A;Ba

##      [,1] [,2] [,3] [,4]
## [1,]    0   80   40    0
```

[6]The dominant eigenvalue is 0.4 in our example.

```
## [2,]   0   0 120  40
## [3,] 100   0   0 100
## [4,]   0   0   0   0
```

The matrix element $w_1 a_{12} = 80$ states that 80 employees are necessary to provide the input of good 1 (wood) which is used in the steel production process by ω_2. The labor inverse $\mathbf{H} = \text{diag}(\mathbf{w})(\mathbf{I} - \boldsymbol{\Pi})^{-1}$ is obtained as

```
H <- diag(as.vector(w))%*%IPi;H
```

```
##       [,1]  [,2]  [,3]   [,4]
## [1,] 2.193 1.754  9.65  0.461
## [2,] 0.263 4.211 13.16  0.855
## [3,] 1.096 0.877 54.82  1.480
## [4,] 0.000 0.000  0.00 10.000
```

The entry $w_3 c_{31} = 1.0965$ states that 1.0965 employees have to work in production unit ω_3 to provide the input necessary to produce one unit (e.g., one cord) wood in production unit 1. The sectoral employment multipliers $\mathbf{w}'(\mathbf{I} - \boldsymbol{\Pi})^{-1}$ are obtained as

$$\begin{aligned} &\left[\sum_i^n w_i c_{i1}, \ldots, \sum_i^n w_i c_{in}\right] \\ &= \left[3.552\,6,\ 6.842\,1,\ 77.631\,6,\ 12.796\,1\right]. \end{aligned} \tag{2.22}$$

```
t(w)%*%IPi
```

```
##      [,1] [,2] [,3] [,4]
## [1,] 3.55 6.84 77.6 12.8
```

The multiplier $\sum_i^n w_i c_{i3} = 77.631\,6$ states the number of employees that are necessary to facilitate the production of one unit of output c_3 in sector 3 available for final demand.

2.2 Input–Output Tables

2.2.1 Aggregated Tables for Germany

The statistical office of Germany provides input–output tables since the 1970s. Contrary to the empirical analysis of single tables (years), the analysis of longer time periods is hampered by incomplete harmonization of the tables across time. The researcher faces the problem of continuous changes in the level of aggregation, the definition of sectors and product groups, and the treatment of trade and banking services.

In recent years, the number of sectors increased from 59 to 71 and the German input–output table for 2007 contains 74 sectors. Somewhat harmonized tables are available for the time period 1991–2007. Nevertheless, not for all years detailed information of all components of value added is provided by the statistical office. In the following, we first discuss the table for 2007 and then analyze the major changes between 1991 and 2007 using both input–output tables. Throughout, we use highly aggregated versions of the original table.

We define 12 broad sectors (production units) and three different components (columns) of final demand (13 consumption, 14 investment, 15 export). The 12 sectors (index $j = 1, \ldots, n$), which correspond to 12 kinds of goods (index $i = 1, \ldots, n$), are denoted as: 1 agriculture, 2 energy, 3 chemistry, 4 metal, 5 machines, 6 textiles, 7 food, 8 construction, 9 trade, 10 renting, 11 social, 12 services. Below the input matrix **A** additional rows provide information on imports, taxes, and components of value added: 13 import, 14 tax, 15 due, 16 wage, 17 depreciation, 18 profit. For further calculations we use the table for 2007 from Statistisches Bundesamt (2010). Tables 2.1, 2.2, and 2.3 contain values in current prices in million Euro.

2.2.2 Some Identities

If we add for unit j (column) the 12 entries in row 1 to 12, we obtain the intermediate consumption of unit j. Adding the tax on products paid minus subsidies on products received (row 14), we obtain intermediate consumption at purchaser's prices.

Table 2.1 Aggregated A-Table 2007 (1). Statistisches Bundesamt (2010). Published with the kind permission of © Statistisches Bundesamt 2010. All rights reserved

	1 agric.	2 energy	3 chem.	4 metal	5 machines	6 textiles
1 agriculture	8319	47	21	0	0	2325
2 energy	1009	26922	11487	6057	5389	4311
3 chemistry	2817	1269	90268	6957	23019	7959
4 metal	357	1621	2410	91327	58571	1490
5 machines	1065	4536	4276	4352	185217	1483
6 textiles	183	350	3775	3091	7489	33713
7 food	3707	0	1028	0	0	3
8 construction	273	1439	893	803	1358	607
9 trade	3442	5037	17540	13972	39688	13990
10 renting	8181	10736	33951	14163	72598	19165
11 social	728	582	3569	1329	1771	1181
12 services	306	5095	2008	861	3980	3852
13 import	4793	15837	100544	60712	154736	31583
14 tax	1235	1501	1951	1575	3639	1626
15 due	−6238	−121	1867	1776	4187	1273
16 wage	8161	19227	49022	43188	157911	34614
17 depreciation	7409	13254	12637	7375	24379	8397
18 profit	10638	16503	18389	17871	38153	13672

Table 2.2 Aggregated A-Table 2007 (2). Statistisches Bundesamt (2010). Published with the kind permission of © Statistisches Bundesamt 2010. All rights reserved

	7 food	8 const.	9 trade	10 renting	11 social	12 services
1 agriculture	23243	0	378	681	358	1011
2 energy	3133	1724	7256	2689	3773	3093
3 chemistry	3581	23114	11111	1836	4103	1750
4 metal	748	9372	2180	351	605	652
5 machines	1139	9665	10952	1719	2366	4862
6 textiles	2426	5065	7302	7386	3469	2518
7 food	24209	7	6811	17	3928	968
8 construction	433	7344	3806	20795	3858	3845
9 trade	16031	12547	137055	10355	11695	10504
10 renting	19182	32295	108746	240705	30073	25129
11 social	924	756	6173	7920	19226	4271
12 services	1247	1695	6794	18794	4031	23720
13 import	22299	17104	52954	39375	13894	11947
14 tax	2247	1626	11771	14998	12079	11205
15 due	532	868	11593	18057	−4692	−832
16 wage	21928	53157	248909	209975	184973	149365
17 depreciation	5206	4762	46495	160112	34867	34687
18 profit	5905	31633	92091	281249	48113	34073

Table 2.3 Aggregated A-Table 2007 (3). Statistisches Bundesamt (2010). Published with the kind permission of © Statistisches Bundesamt 2010. All rights reserved

	13 cons.	14 inv.	15 export
1 agriculture	9352	5520	5130
2 energy	35048	462	11482
3 chemistry	34404	−9279	152727
4 metal	2860	3705	99160
5 machines	47155	103179	400119
6 textiles	40610	660	63207
7 food	75667	3058	35010
8 construction	3594	163517	169
9 trade	349559	19660	111302
10 renting	326202	36022	59866
11 social	328289	0	0
12 services	243814	4603	1968
13 import	124603	69317	176160
14 tax	144774	5116	−150
15 due			
16 wage			
17 depreciation			
18 profit			

Adding up other taxes on production less subsidies on production (due, row 15), wages (16), depreciation (17), and profits (18) results in the value added. The sum of intermediate consumption at purchaser's prices and value added is output.

If we look at one good (row *i*) and add up the 12 entries in columns 1 to 12, we obtain the value of the amount of good *i* that is used as intermediate input in the production units. Summing up the value of consumption (column 13) and investment (14) results in the final domestic demand for good *i*. Adding further exports (15) leads to final demand for good *i*.

The value of the amount of good *i* used as intermediate input and the final demand for good *i* add up to the total use of good *i* which, by definition, equals output of unit *i*.

2.3 Analyzing German Input–Output Tables 1991 and 2007

2.3.1 Handling Input–Output Tables

The following analysis is based on input–output tables for Germany 1991 and 2007. Throughout we only analyze highly aggregated tables with 12 sectors/goods. While there have been attempts to provide input–output tables in constant prices of a base year, the procedure is rather problematic, as meaningless results are likely to occur. For example if the ratio of input to output prices changes, producers react with their input and output mix. Combining quantities that have been chosen according to a specific price scheme with outdated price schemes can easily result in meaningless figures, e.g., a positive value added in current prices turning negative in constant prices. Therefore, we adjust input–output tables only for overall price changes leaving the price ratios unaltered. To measure overall price changes, we use the price index of final domestic use. Choosing 2007 as base year with price index set to the value 100, we inflate the input–output table from 1991 by the factor 1.243 556. Note that the choice of the base year does neither effect real growth rates nor sector shares.

We start with reading the input–output table for 2007 into an R object. Using the function `read.table()`, we have to specify the name of the object that will contain the table and the file-name that is to be loaded. Additionally, we specify that ";" separates fields (delimiter) in the file and that "," is the character used in the file for decimal points. We divide all input–output and employment data by 1000, which leaves input–output data in billions and employment data in millions.

```
d1 <- read.table("io_2007.csv", sep=";", dec=",")/1000
```

A shortcut to this command is the implemented function `read.csv2()` that uses the mentioned options by default:

```
d1 <- read.csv2("io_2007.csv")/1000
```

In the same way, we read the input–output table for 1991 (d0), the employment data for 2007 (e1) and for 1991 (e0).

```
d0.org <- read.csv2("io_1991.csv")/1000
e1 <- read.csv2("et_2007.csv")/1000
e0 <- read.csv2("et_1991.csv")/1000
```

To allow comparisons of the tables for 1991 and 2007, we inflate the input–output table of 1991 by the overall price change that has taken place between 1991 and 2007 using the index of final domestic demand. Note that inflating the table from 1991 towards the overall price level of 2007 does not alter the relative prices of 1991. Hence, differences between the two tables (1991 and 2007) will be the combined result of changes in quantities and of changes in relative prices.

```
pfile <- read.csv2("pliv.csv")
p <- pfile$pliv
dp <- p[pfile$year==2007]/p[pfile$year==1991];dp

## [1] 1.24

d0 <- d0.org*dp
```

The input tables (d0, d1) and the employment data (e0, e1) are stored in the working directory as dataframes by default. To ease the handling of the objects, we use matrices and vectors instead:

```
d0 <- as.matrix(d0)
d1 <- as.matrix(d1)
e0 <- as.vector(as.matrix(e0))
e1 <- as.vector(as.matrix(e1))
```

Using brackets, we can pick elements of the io-table, e.g., the value of chemical products (good $i = 3$) used as input in agriculture (unit $j = 1$):

```
d1[3,1]

## [1] 2.82
```

The identities discussed above can be used to obtain (gross) output and value added:

```
# gross output
x1 <- colSums(d1[,1:12]);x1

##  agriculture       energy    chemistry        metal
##         56.4        123.8        355.6        275.4
##     machines     textiles         food construction
##        782.1        181.2        154.4        212.7
##        trade      renting       social     services
##        772.4       1037.0        376.7        322.8

# value added
va1 <- colSums(d1[c(15:18),1:12]);va1
```

```
## agriculture        energy     chemistry        metal
##        20.0          48.9          81.9         70.2
##    machines      textiles          food construction
##       224.6          58.0          33.6         90.4
##       trade       renting        social     services
##       399.1         669.4         263.3        217.3
```

In the same way, we calculate x0 and va0 with the 1991 table.

```
x0 <- colSums(d0[,1:12]);x0

## agriculture        energy     chemistry        metal
##        52.9         100.9         226.7        156.8
##    machines      textiles          food construction
##       489.2         192.5         149.1        217.5
##       trade       renting        social     services
##       588.7         639.7         245.2        257.3

va0 <- colSums(d0[c(15:18),1:12]);va0

## agriculture        energy     chemistry        metal
##        24.8          49.5          81.8         61.3
##    machines      textiles          food construction
##       197.7          73.7          37.0        106.0
##       trade       renting        social     services
##       327.9         374.5         174.4        182.0
```

2.3.2 Some Empirical Results

Sectoral shares in output and in value added are obtained as

$$\frac{x_{t_0j}}{\sum_{k=1}^{n} x_{t_0k}} \quad \text{and} \quad \frac{va_{t_0j}}{\sum_{k=1}^{n} va_{t_0k}}. \tag{2.23}$$

The sectoral shares x1s in output can be calculated dividing x1 element by element by the sum of x1:

```
x1s <- x1/sum(x1)
```

To compare the sectoral shares in output in 1991 and 2007, we bind the vectors with shares in 1991 and 2007 and display the shares using a barplot. As labels, we use the sector names given in the input–output tables. See Fig. 2.1.

```
x0s <- x0/sum(x0)
tab01 <- rbind(x0s,x1s)
barplot1 <- barplot(tab01, main="", ylim=c(0,0.25),
                    ylab="Share of gross output (in %)",
                    col=c("grey60","grey90"),
                    beside=TRUE, xaxt="n")
```

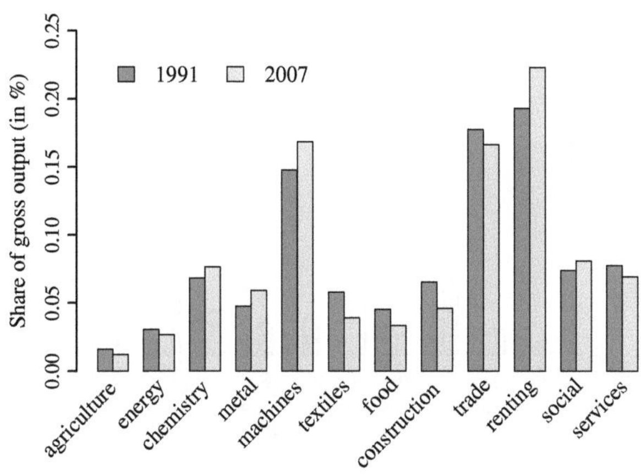

Fig. 2.1 Sectoral shares in gross output by sector for 1991 and 2007

```
text(apply(barplot1, 2, mean),0,
     labels=rownames(d1)[1:12],srt=45,
     adj=c(1.1,1.1),xpd=TRUE,cex=1.1)
legend(1, 0.24, c("1991", "2007"),
       fill=c("grey60","grey90"),
       bty="n", ncol=2, cex=1.1)
```

We find renting, trade, and machines to be the biggest sectors. Five sectors have increased their share in overall output at the expense of the remaining seven sectors: chemistry, metal, machines, renting, and social.

Next, we compare sectoral shares in 2007 in output and value added. Sectoral value added can be obtained by summing up taxes on production less subsidies on production (row 14), dues (other taxes) (15), wages (16), depreciation (17), and profits (18). Of course, subtracting intermediate consumption, which is obtained by summing up all kinds of inputs (rows 1 to 12), also results in value added. See Fig. 2.2.

```
vals <- val/sum(val)
tab02 <- rbind(xls,vals)
barplot2 <- barplot(tab02, main="", ylim=c(0,0.35),
                    ylab="Share of output/va (in %)",
                    col=c("grey60","grey90"),
                    beside=TRUE, xaxt="n")
text(apply(barplot2, 2, mean),0,
     labels=rownames(d1)[1:12],srt=45,
     adj=c(1.1,1.1),xpd=TRUE,cex=1.1)
legend(1, 0.34, c("output", "value added"),
       fill=c("grey60","grey90"), bty="n",
       ncol=2,cex=1.1)
```

We find renting to be the most important sector according to its share in output as well as in value added. It is remarkable that about one-third of total value added

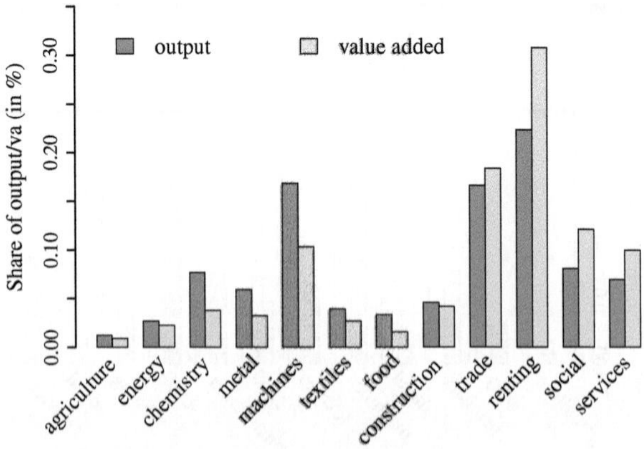

Fig. 2.2 Sectoral shares in output and value added (va) for 2007

is earned in renting. Note that the treatment of rents as income included in value added is a questionable convention and implies that value added must not be seen as a measure of production. If income would be distributed more equally and people would own their homes, there would be no contribution of rents to total value added. The second biggest sector is trade. Note that shares in gross production and in value added are very alike. This is also the effect of a convention in national accounting: the value of purchased goods is not counted as input in the trade sector. We find strong differences between shares in output and value added for chemistry and machines. Differences in the shares for output and value added imply differences in the ratio of value added to output (va_j/x_j). We find the highest value added to output ratios in renting, social, and services.

```
round(va1/x1,2)

##  agriculture       energy   chemistry        metal
##         0.35         0.39        0.23         0.25
##     machines     textiles        food construction
##         0.29         0.32        0.22         0.43
##        trade      renting      social     services
##         0.52         0.65        0.70         0.67
```

Comparing the sectoral shares in value added in 1991 and 2007 reveals that only two sectors have gained shares: renting and social. The remaining ten sectors have lower shares in 2007 compared to 1991. This fact implies that growth in value added has been strongest in renting and social.

```
va0s <- va0/sum(va0)
round(rbind(va0s,va1s),3)

##      agriculture energy chemistry metal machines
## va0s       0.015  0.029     0.048 0.036    0.117
## va1s       0.009  0.022     0.038 0.032    0.103
```

```
##      textiles  food construction trade renting social
## va0s    0.044 0.022       0.063 0.194   0.222  0.103
## va1s    0.027 0.015       0.042 0.183   0.308  0.121
##      services
## va0s    0.108
## va1s    0.100
```

Next, we compare the overall growth rate (percent) in real value added, that is using the price level of 2007 also for the table of year 1991:

```
round(va1/va0*100-100,2)
```

```
##  agriculture       energy    chemistry        metal
##       -19.48        -1.33         0.10        14.52
##     machines     textiles         food construction
##        13.62       -21.34        -9.26       -14.70
##        trade      renting       social     services
##        21.72        78.75        50.99        19.38
```

We find that five sectors show a decline in real value added. Strong increases are observed for renting and social.

Generally, productivity is obtained by dividing output through input. To compare sectoral labor productivity in 1991 and 2007, we have to decide about the concepts of measuring "output" and "input". Here, we use value added to measure output and labor input is measured using the number of employees. Note that we use prices of 2007 in both tables, therefore, changes in productivity include no effects of overall inflation but changes in relative prices. An increase in labor productivity can be caused by sectoral relative price increases or by increases in quantity. See Fig. 2.3.

```
lp0 <- colSums(d0[15:18,1:12])/e0
lp1 <- colSums(d1[15:18,1:12])/e1
tab03 <- rbind(lp0,lp1);tab03
```

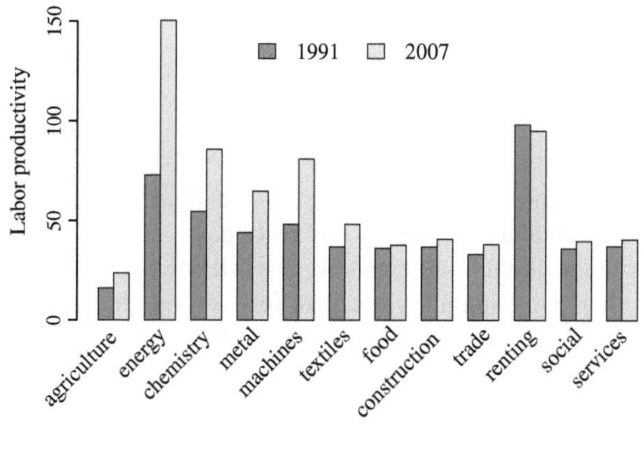

Fig. 2.3 Value added productivity of labor 1991 and 2007 (in th. euro)

```
##     agriculture energy chemistry metal machines
## lp0        16.0   72.8      54.6  44.1     48.2
## lp1        23.8  150.3      85.6  64.8     80.9
##     textiles food construction trade renting social
## lp0     37.0 36.3         37.1  33.3    98.4   36.1
## lp1     48.2 37.9         40.8  38.3    94.9   39.8
##     services
## lp0     37.3
## lp1     40.7
```

```
barplot3 <- barplot(tab03, main="", ylim=c(0,170),
                    ylab="Labor productivity",
                    col=c("grey60","grey90"),
                    beside=TRUE, xaxt="n")
text(apply(barplot3, 2, mean),0,
     labels=rownames(d1)[1:12],srt=45,
     adj=c(1.1,1.1),xpd=TRUE,cex=1.1)
legend(10, 150, c("1991", "2007"),
       fill=c("grey60","grey90"),
       bty="n",ncol=2,cex=1.1)
```

We observe that in 2007 labor productivity is outstandingly high in the energy sector. In this sector, we also observe the strongest increase in labor productivity. Noteworthy increases also took place in chemistry, metal, and machines.

Labor productivity is strongly influenced by capital intensity. Capital intensity is obtained by dividing a measure of capital input through a measure of labor input. We would prefer quantities or surrogates of quantities (capital stock, that is quantities valued with base year (constant) prices) for labor and capital use. Unfortunately, the German statistical office does not provide capital stock data disaggregated for twelve (or more) sectors. We obtain the following figures of the ratio of depreciation to wages as an approximate measure of capital intensity.

```
round(d1[17,1:12]/d1[16,1:12],2)

##  agriculture       energy    chemistry        metal
##         0.91         0.69         0.26         0.17
##     machines     textiles         food construction
##         0.15         0.24         0.24         0.09
##        trade      renting       social     services
##         0.19         0.76         0.19         0.23
```

Interestingly, the highest capital intensity is found in agriculture. High capital intensities are also obtained for energy and renting.

Sectoral (column) input coefficients π_{ij} provide information on the production process of sector j. As a summary measure for sectoral changes, we compare the sum of absolute changes in coefficients between 1991 and 2007 given by

$$\sum_{i=1}^{n} \left| \pi_{2007,ij} - \pi_{1991,ij} \right| .$$

To obtain these sector specific summary measures, we make use of the recycling rule implemented in R. A short object is recycled to meet the length of the longer object. To make yourself comfortable with this rule, you can try the following commands:

```
a <- 1:9
b <- 1:3
A <- matrix(a,3,3)
a+b

## [1]  2  4  6  5  7  9  8 10 12

A/b

##      [,1] [,2] [,3]
## [1,]    1  4.0    7
## [2,]    1  2.5    4
## [3,]    1  2.0    3

t(A)/b

##      [,1] [,2] [,3]
## [1,] 1.00 2.00    3
## [2,] 2.00 2.50    3
## [3,] 2.33 2.67    3

t(t(A)/b)

##      [,1] [,2] [,3]
## [1,]    1  2.0 2.33
## [2,]    2  2.5 2.67
## [3,]    3  3.0 3.00
```

In our empirical example, we want to divide column j by the output of this sector x_j. We use A for the matrix of input in Euro and P for the matrix of input coefficients π_{ij}.

```
# 1991
A0 <- d0[1:12,1:12]
x0 <- colSums(d0[,1:12])
P0 <- t(t(A0)/x0)
# 2007
A1 <- d1[1:12,1:12]
x1 <- colSums(d1[,1:12])
P1 <- t(t(A1)/x1)
```

Sectoral changes in the input structure can now easily be obtained:

```
round(colSums(abs(P1-P0)),3)

## agriculture       energy    chemistry        metal
##       0.264        0.084        0.171        0.173
##    machines     textiles         food construction
##       0.135        0.089        0.154        0.112
##       trade      renting       social     services
##       0.099        0.087        0.107        0.095
```

We find that in sector agriculture, the input structure has changed dramatically and more than in any other sector. In chemistry, metal, and food sectors, we also observe a considerable change in the input structure. Little change is observed in energy, textile, and renting.

Next, we want to analyze the total amount of induced production through additional sectoral demand. Note that additional demand for product j causes additional production in sector j. To increase its output sector j has to use inputs according to its input structure. These inputs will be produced in all sectors whose products are used as inputs in sector j. But these inputs will have to be produced again by means of other inputs. We can imagine this process ad infinitum. Therefore, the output $\mathbf{x}$ that must be produced to allow final demand $\mathbf{y}$ will exceed that amount of final demand considerably. Mathematically, the relation between $\mathbf{y}$ and total output $\mathbf{x}$ is obtained using equation

$$\mathbf{x} = (\mathbf{I} - \boldsymbol{\Pi})^{-1}\mathbf{y}. \tag{2.24}$$

Note that by setting $dy_j = 1$, we simply pick a single column j of the Leontief-inverse $(\mathbf{I}-\boldsymbol{\Pi})^{-1}$. If we add up the coefficients c_{ij} in column j, we obtain the amount of production that is necessary to allow one unit final demand for good j. The vector of the sums of columns therefore shows the production stimulus that results from demand for the different kinds of goods. In R, we first calculate the Leontief-inverse matrices for both periods and sum up the columns. Results are presented by means of a barplot (see Fig. 2.4):

```
n <- 12
leo0 <- solve(diag(n)-P0)
leo1 <- solve(diag(n)-P1)
tab04 <- rbind(colSums(leo0),colSums(leo1))
barplot4 <- barplot(tab04-1.3, main="",  ylim=c(0,1),
```

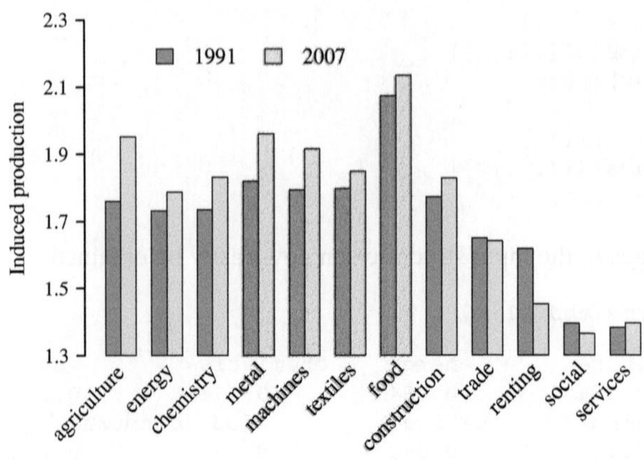

Fig. 2.4 Induced production by one unit of final demand

```
                    ylab="Induced production",
                    col=c("grey60","grey90"),beside=T,
                    las=2, axes=F, xaxt="n")
text(apply(barplot4, 2, mean),0,
    labels=rownames(d1)[1:12],srt=45,
    adj=c(1.1,1.1),xpd=TRUE,cex=1.1)
axis(2, at=seq(0,1,0.2), labels=seq(0,1,0.2)+1.3, las=2)
legend(3, 0.98, c("1991", "2007"),
      fill=c("grey60","grey90"),
      bty="n",ncol=2,cex=1.1)
```

We find that in the food sector, an additional demand for one Euro causes production to increase by more than two Euro. We observe low values of induced production in the social and the service sectors.

While the coefficients of the Leontief-inverse in one column show the total amount of induced production in the input producing sectors, we can express these induced production in labor. We use the average amount of labor per unit output $b_i/x_i = w_i$ used in each of the n sectors to express induced additional input production (output of sector i induced by final demand for good j) c_{ij} in sector i in labor $w_i \cdot c_{ij}$. Using matrix notation, the employment effect is obtained using equation

$$\mathbf{w}'(\mathbf{I} - \boldsymbol{\Pi})^{-1}\mathbf{y}. \tag{2.25}$$

As we have measured the data in the input–output table in billions and the employment data in millions, the results show the number of additional employees per million additional final demand for good j. For 2007, we obtain the following employment effects, often called sectoral employment multipliers:

```
w0 <- e0/x0*1000;w0

##  agriculture       energy    chemistry        metal
##        29.30         6.74         6.61         8.87
##     machines     textiles         food construction
##         8.39        10.33         6.83        13.16
##        trade      renting       social     services
##        16.73         5.95        19.71        18.95

w1 <- e1/x1*1000;w1

##  agriculture       energy    chemistry        metal
##        14.86         2.62         2.69         3.94
##     machines     textiles         food construction
##         3.55         6.63         5.73        10.42
##        trade      renting       social     services
##        13.51         6.80        17.56        16.55

H0 <- diag(w0)%*%leo0
H1 <- diag(w1)%*%leo1
m0 <- colSums(H0);m0 # or: t(w0)%*%leo0
```

```
##  agriculture         energy     chemistry          metal
##         37.7           13.5          13.1           16.7
##     machines       textiles          food construction
##         15.7           18.4          21.9           20.1
##        trade        renting        social       services
##         23.8           11.0          23.7           23.4
```

```
m1 <- colSums(H1);m1 # or: t(w1)%*%leo1
```

```
##  agriculture         energy     chemistry          metal
##        23.30           7.69          7.53           9.46
##     machines       textiles          food construction
##         8.90          13.30         16.06          15.96
##        trade        renting        social       services
##        19.47          10.43         21.00          20.46
```

Employment is stimulated most by additional final demand for goods of the sectors agriculture, trade, social, and services. Very little employment effects are observed for the machines sector, where the expensive stimuli ("Abwrackpraemie", governmental subsidy for buying a new car when wrecking your old car) in the last banking induced crisis were focused on. See Fig. 2.5.

```
tab005 <- cbind(m0,m1);tab005
```

```
##                m0    m1
## agriculture  37.7 23.30
## energy       13.5  7.69
## chemistry    13.1  7.53
## metal        16.7  9.46
## machines     15.7  8.90
## textiles     18.4 13.30
## food         21.9 16.06
## construction 20.1 15.96
```

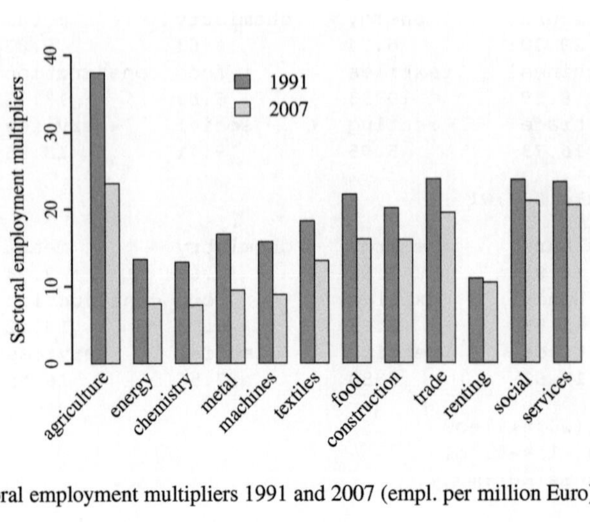

Fig. 2.5 Sectoral employment multipliers 1991 and 2007 (empl. per million Euro)

```
## trade          23.8 19.47
## renting        11.0 10.43
## social         23.7 21.00
## services       23.4 20.46
```

```
barplot5 <- barplot(t(tab005), main="", ylim=c(0,40),
                ylab="Sectoral employment multipliers",
                col=c("grey60","grey90"),
                beside=T, xaxt="n")
text(apply(barplot5, 2, mean),0,
     labels=rownames(d1)[1:12],srt=45,
     adj=c(1.1,1.1),xpd=TRUE,cex=1.1)
legend(10,40, c("1991", "2007"),
       fill=c("grey60","grey90"),
       bty="n",cex=1.1)
```

2.4 Recommended Readings

The theoretical model is mainly based on the works of

- Schwartz J (1961) Lectures on the mathematical method in analytical economics. Mathematics and its applications. Gordon and Breach, New York

and

- Sraffa P (1963) Production of commodities by means of commodities. Prelude to a critique of economic theory. Cambridge University Press, Cambridge.

A reference monograph on linear production models is

- Gale D (1989) The theory of linear economic models. University of Chicago Press, Chicago.

A recommendable modern textbook focusing on theoretical aspects of input–output analysis is

- ten Raa T (2005) The economics of input-output analysis. Cambridge University Press, Cambridge.

Many extensions of the input–output model as well as many practical problems are discussed in

- Miller R, Blair P (2009) Input-output analysis: foundations and extensions. Cambridge University Press, Cambridge.

2.5 Exercises

The data files `io_1991.csv`, `io_2007.csv`, `et_1991.csv`, and `et_2007.csv` contain highly aggregated input–output tables and employment data for Germany 1991 and 2007. The file `pliv.csv` contains the price index of final domestic demand (Consumption and Investment).

1. The data structure
   a. Load the data into objects `d0.org`, `d1`, `e0`, `e1`, `pliv`.
   b. Calculate the price increase between 1991 and 2007.
   c. Convert your table `d0.org` (1991) into the table `d0` inflated towards the price level of 2007.
2. Some graphics
   a. Make a barplot showing the gross production in 1991 and 2007 by sectors.
   b. Make a barplot displaying the percentage changes in gross production by sectors between 1991 and 2007.
   c. Make a barplot comparing the sectoral shares in value added and gross production in 2007. What do differences in shares imply?
   d. Make a barplot comparing the value added productivity of labor in 1991 and 2007 by sectors.
3. Sectoral multipliers
   a. Calculate the sectoral employment multipliers for 1991 and 2007.
   b. Did multipliers overall increase or decrease since 1991?
   c. Display the results graphically.
4. Growth in real wages and real profits
   a. Calculate the ratio of profits to wages by sector in 1991 and 2007 and display the results graphically.
   b. Calculate the sectoral growth rates of wages and profits and display the results graphically.

References

Behr A, Rohwer G (2013) Wirtschafts- und Bevölkerungsstatistik. UTB GmbH

Gale D (1989) The theory of linear economic models. University of Chicago Press, Chicago

Miller R, Blair P (2009) Input-output analysis: foundations and extensions. Cambridge University Press, Cambridge

ten Raa T (2005) The economics of input-output analysis. Cambridge University Press, Cambridge

Schwartz J (1961) Lectures on the mathematical method in analytical economics. Mathematics and its applications. Gordon and Breach, New York

Sraffa P (1963) Production of commodities by means of commodities. Prelude to a critique of economic theory. Cambridge University Press, Cambridge

Statistisches Bundesamt (2010) Input-output-rechnung, 2007. Volkswirtschaftliche Gesamtrechnungen 18(2), Wiesbaden

Production Functions 3

Abstract

Production functions postulate a functional relation between inputs and output. In the economic literature three production functions seem to play a dominant role: The Cobb–Douglas production function, the constant elasticity of substitution function, and the transcendental logarithmic production function. In this chapter, we provide a short discussion of their theoretical properties. Overlaying the deterministic output with a random error, we show how to generate artificial data according to the production functions and demonstrate how to estimate the parameters of the different production functions.

3.1 Introduction

Production functions formulate algebraically a relationship between inputs x_i $(i = 1, \ldots, r)$ and output y:

$$y = f(x_1, \ldots, x_r). \tag{3.1}$$

An interesting historical account of the production function in economics is provided by Mishra (2007). Early concepts of the production function date back towards von Thünen (1910) and Wicksteed (1894). Empirically, most firms produce technically not fully efficient. That is they could either maintain their output and reduce their inputs or could increase their output using the given amount of inputs. Despite this fact, in this chapter we assume technical efficient production throughout.

Furthermore, throughout this chapter (and the next) we assume that a firm produces only one homogeneous output and uses only one kind of homogeneous capital

A. Behr, *Production and Efficiency Analysis with R*,
DOI 10.1007/978-3-319-20502-1_3

as input. Both, homogeneous output and homogeneous capital could be measured in suitable quantities. Thereby, we circumvent all problems of aggregation. When using empirical data (in contrast to generated data by means of a random number generator) one has to be aware that one faces severe problems of valuation and aggregation.

In this chapter, we discuss three production functions which are most often used in empirical analysis. The Cobb–Douglas (CD) production function is the most prominent production function in economics. Its main charm is the easy mathematical handling which made the function the basic model of production in standard micro- and macroeconomic textbooks. The literature on the Cobb–Douglas function is tremendous but also highly controversial. Theoretically, it has been shown that due to the possibility of reswitching, an aggregate production function is meaningless.[1] Robinson (1953) called the standard textbook exposition "miseducation". Furthermore, it is often argued that the Cobb–Douglas production function is merely a bookkeeping identity and of no empirical relevance [cf. Shaikh, 1974].

The constant elasticity of substitution (CES) production function encompasses the Cobb–Douglas function allowing the elasticity of substitution between different inputs to deviate from 1. While theoretically superior, the CD is just a special case of the CES, estimation of the CES production function is sometimes difficult because of its nonlinearity. This can result in rather flat shapes of the likelihood and occurrences of multiple optima. As this problem easily occurs even in the case of simulations using the correct model, the empirical usefulness of the CES production function is very limited.

The transcendental logarithmic production function (TL) is an approximation to complex nonlinear production relations and easy to handle. Nowadays, the TL production function can be seen as the standard production function in empirical analysis due to its flexibility and linearity in parameters which strongly simplifies estimation. Empirically, the estimated coefficients of the TL rather often imply theoretically implausible properties of the production process and an explicit interpretation of coefficients is often avoided.

In the following, we try to give a very concise presentation of the production functions and their properties and focus on the estimation using the R environment.

[1] To obtain the relative price of labor and capital the amounts of labor and capital are taken as given. But the aggregation of capital itself already rests on relative prices of capital. The famous Cambridge controversy focused on the question whether capital intensity would monotonically increase with a decreasing relative price of capital (no reswitching). During the debate it has been clarified that reswitching cannot be ruled out: "We wish to make it clear for the record that the nonreswitching theorem associated with us is definitely false. We are grateful to Dr. Pasinetti ..." Levhari and Samuelson (1966).

3.2 The Cobb–Douglas Function

The Cobb–Douglas production function (CD) dates back to von Thünen (1850), Wicksell (1923), and Wicksell (1934) but is attributed in the literature to (and named after) Cobb and Douglas (1928).

3.2.1 Properties of the Function

We denote the inputs with x_i $(i = 1, \ldots, r)$ and the output with y

$$y = f(x_1, x_2, \ldots, x_r) = \beta_0 \prod_{i=1}^{r} x_i^{\beta_i}$$
$$\text{with } \beta_0 > 0, 0 < \beta_{i,i\neq 0} < 1 \text{ and } \sum\nolimits_{i=1}^{r} \beta_i = 1. \tag{3.2}$$

The marginal product is positive

$$\frac{\partial y}{\partial x_i} = \beta_0 \beta_i x_i^{\beta_i - 1} \prod_{j=1, j\neq i}^{r} x_j^{\beta_j} = \beta_i x_i^{-1} \beta_0 \prod_{j=1}^{r} x_j^{\beta_j} = \beta_i x_i^{-1} y. \tag{3.3}$$

With factor prices w_i $(i = 1, \ldots, n)$ and output price p, profit maximization results in

$$\frac{\partial y}{\partial x_i} = \frac{w_i}{p}. \tag{3.4}$$

The ratio of the marginal productivity of two factors is referred to as the marginal rate of technical substitution (MRTS)

$$\frac{\partial y/\partial x_i}{\partial y/\partial x_j} = \frac{\beta_i x_i^{-1} y}{\beta_j x_j^{-1} y} = \frac{\beta_i x_j}{\beta_j x_i} = \frac{w_i/p}{w_j/p} = \frac{w_i}{w_j}. \tag{3.5}$$

Second order partial derivatives are

$$\frac{\partial^2 y}{\partial x_i^2} = \frac{\partial(\beta_i x_i^{-1} \beta_0 \prod_{j=1}^{r} x_j^{\beta_j})}{\partial x_i} = \beta_i(\beta_i - 1) x_i^{\beta_i - 2} \beta_0 \prod_{j=1, j\neq i}^{r} x_j^{\beta_j}$$
$$= \beta_i(\beta_i - 1) x_i^{-2} \beta_0 \prod_{j=1}^{r} x_j^{\beta_j} = -\beta_i(1 - \beta_i) x_i^{-2} y < 0 \tag{3.6}$$

and

$$\frac{\partial^2 y}{\partial x_i x_j} = \frac{\partial(\beta_i x_i^{-1}\beta_0 \prod_{k=1}^r x_k^{\beta_k})}{\partial x_j} = \beta_i \beta_j x_i^{-1} x_j^{\beta_j - 1} \beta_0 \prod_{k=1,k\neq j}^{r} x_k^{\beta_k}$$
$$= \beta_i \beta_j x_i^{-1} x_j^{-1} \beta_0 \prod_{k=1}^{r} x_k^{\beta_k} = \beta_i \beta_j x_i^{-1} x_j^{-1} y > 0. \tag{3.7}$$

The Cobb–Douglas function is linearly homogeneous

$$y\theta = \beta_0 \prod_{i=1}^{r} (\theta x_i)^{\beta_i} = \beta_0 \prod_{i=1}^{r} \theta^{\beta_i} x_i^{\beta_i} = \theta^{\sum_{i=1}^{r} \beta_i} \beta_0 \prod_{i=1}^{r} x_i^{\beta_i}. \tag{3.8}$$

If real factor price equals marginal product, the income share equals the exponent of the production factor

$$\frac{\partial y}{\partial x_i} = \frac{w_i}{p} = \beta_i x_i^{-1} y \Rightarrow \frac{x_i w_i}{yp} = \beta_i. \tag{3.9}$$

Euler's theorem states that the total product will be exhausted if the factors are paid according to their marginal product. For the CD the theorem holds, as factor rewards exhaust the product:

$$\sum_{i=1}^{r} \beta_i y = \sum_{i=1}^{r} \frac{\partial y}{\partial x_i} x_i = \sum_{i=1}^{r} \beta_i x_i^{-1} y x_i = y \sum_{i=1}^{r} \beta_i = y. \tag{3.10}$$

The elasticity of substitution defined as the relative change of inputs to relative change of input prices is 1. Because of

$$\frac{(x_i w_i)/yp}{(x_j w_j)/yp} = \frac{x_i/x_j}{w_j/w_i} = \frac{\beta_i}{\beta_j} \Leftrightarrow \frac{x_i}{x_j} = \frac{w_j}{w_i}\frac{\beta_i}{\beta_j} \Rightarrow \frac{\partial(x_i/x_j)}{\partial(w_j/w_i)} = \frac{\beta_i}{\beta_j} \tag{3.11}$$

we find

$$\sigma = \frac{\frac{\partial(x_i/x_j)}{x_i/x_j}}{\frac{\partial(w_j/w_i)}{w_j/w_i}} = \frac{\partial(x_i/x_j)}{\partial(w_j/w_i)} \cdot \frac{w_j/w_i}{x_i/x_j} = \frac{\beta_i}{\beta_j}\frac{\beta_j}{\beta_i} = 1. \tag{3.12}$$

3.2.2 Estimating the CD with R

Linearization through taking logs facilitates the estimation of the Cobb–Douglas production function. We generate artificial data according to the following production function

$$y = f(x)e^u = e^{\beta_0} x_1^{\beta_1} x_2^{\beta_2} e^u \qquad \text{with } \beta_1 + \beta_2 = 1 \tag{3.13a}$$

or use the log model

$$\ln y = \beta_0 + \beta_1 \ln x_1 + \beta_2 \ln x_2 + u. \tag{3.13b}$$

In R, the data can be generated according to the model as:

```
n <- 100
b0 <- 1
b1 <- 0.7
b2 <- 0.3
set.seed(123)
# lin-lin-model
x1 <- exp(runif(n))
x2 <- exp(runif(n))
u <- rnorm(n,sd=0.2)
y <- exp(b0)*x1^b1*x2^b2*exp(u)
# create logs
lx1 <- log(x1)
lx2 <- log(x2)
ly <- log(y)
```

We display the data using a 3D-scatterplot in Fig. 3.1.

```
library(scatterplot3d)
scatterplot3d(x1, x2, y, main="",
              xlab=expression(x[1]),
              ylab=expression(x[2]),
              zlab="y",
              color="darkgrey", pch=20, type="h")
```

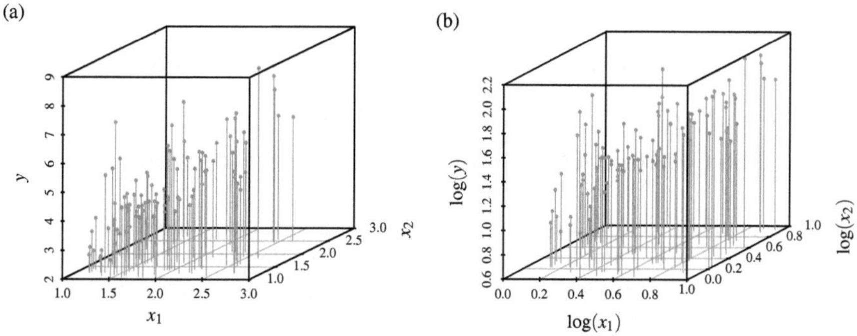

Fig. 3.1 Cobb–Douglas production function. (**a**) Original data. (**b**) Data in logs

```
scatterplot3d(lx1, lx2, ly, main="",
    xlab=expression(paste("log(", x[1], ")")),
    ylab=expression(paste("log(", x[2], ")")),
    zlab="log(y)",
    color="darkgrey", pch=20, type="h")
```

For estimating the regression, we use the function `lm()`.

```
reg1 <- lm(log(y)~log(x1)+log(x2))
coef(reg1)

## (Intercept)      log(x1)      log(x2)
##      0.9718       0.7195       0.2941
```

The command `summary(reg1)` would provide more details of the estimated regression.

We next test a simple linear restriction based on the following test statistic

$$\frac{1}{q}(R\hat{\beta} - r)' \left[\hat{\sigma}^2 R(X'X)^{-1}R'\right]^{-1} (R\hat{\beta} - r) \sim F(q, n-k). \tag{3.14}$$

where X denotes the model matrix, R the matrix picking the relevant parts of the parameter vector β, and r contains the constants of the hypothesis. The hypothesis we want to test is $H_0 : \beta_1 = 0.65, \beta_2 = 0.35$. To test this hypothesis we need the estimated variance–covariance matrix of the regression parameters and we have to define r and R.

```
b <- matrix(coef(reg1))
v <- vcov(reg1)
R <- rbind(c(0,1,0),c(0,0,1))
r <- matrix(c(0.65,0.35))
q <- nrow(R)
n_k <- length(y)-length(coef(reg1))
Fs <- 1/q*t(R%*%b-r)%*%solve(R%*%v%*%t(R))%*%(R%*%b-r)
1-pf(Fs,q,n_k)

##        [,1]
## [1,] 0.4276
```

We obtain a p-value above the 5 % level and would not reject the false hypothesis.

3.3 The Constant Elasticity of Substitution Function

The constant elasticity of substitution function (CES) was introduced by Arrow et al (1961). In the Cobb–Douglas production function the elasticity of substitution has value 1. For example if the price ratio of labor to capital increases by one percent, firms will increase the ratio of capital to labor by one percent. The production function formulated by Arrow et al. permitted the elasticity of substitution to lie between zero and infinity. Whereas the value of the elasticity can deviate from one, the CES shares the property with the CD that the elasticity is fixed irrespective of the quantity of output produced and of the quantity of inputs (capital and labor) used in the production process.

3.3.1 Properties of the Function

While the elasticity of substitution is constrained to the value 1 for the Cobb–Douglas production function, the constant elasticity of substitution (CES) production function only restricts σ to be the same for all pairs of inputs:

$$\begin{aligned} &y = \beta_0\left(\sum_{i=1}^{r}\beta_i x_i^{-\rho}\right)^{-\frac{1}{\rho}} \\ &\text{where} \\ &\beta_0\text{: efficiency parameter with } \beta_0 > 0, \\ &\beta_i\text{: distribution parameter with } 0 < \beta_i < 1 \text{ and} \\ &\rho\text{: substitution parameter with } -1 < \rho < \infty. \end{aligned} \tag{3.15}$$

The marginal product of factor x_i is

$$\begin{aligned} \frac{\partial y}{\partial x_i} &= -\frac{1}{\rho}\beta_0\left(\sum_{i=1}^{r}\beta_i x_i^{-\rho}\right)^{-\frac{1}{\rho}-1}(-\rho\beta_i x_i^{-\rho-1}) \\ &= \beta_0\left(\sum_{i=1}^{r}\beta_i x_i^{-\rho}\right)^{-\frac{1}{\rho}}\left(\beta_0\left(\sum_{i=1}^{r}\beta_i x_i^{-\rho}\right)^{-\frac{1}{\rho}}\right)^{\rho}\beta_0^{-\rho}\beta_i x_i^{-(1+\rho)} \\ &= y^{1+\rho}\beta_0^{-\rho}\beta_i x_i^{-(1+\rho)}. \end{aligned} \tag{3.16}$$

Profit maximization requires that marginal product equals real factor price

$$\frac{\partial y}{\partial x_i} = y^{1+\rho}\beta_0^{-\rho}\beta_i x_i^{-(1+\rho)} = \frac{w_i}{p}. \tag{3.17}$$

The marginal rate of technical substitution is

$$\frac{\partial y/\partial x_i}{\partial y/\partial x_j} = \frac{y^{1+\rho}\beta_0^{-\rho}\beta_i x_i^{-(1+\rho)}}{y^{1+\rho}\beta_0^{-\rho}\beta_j x_j^{-(1+\rho)}} = \frac{\beta_i}{\beta_j}\left(\frac{x_j}{x_i}\right)^{(1+\rho)} = \frac{w_i}{w_j}. \tag{3.18}$$

Second order derivatives are

$$\frac{\partial^2 y}{\partial x_i^2} = \frac{1+\rho}{y}\frac{\partial y}{\partial x_i}\frac{(\partial y/\partial x_i)x_i - y}{x_i} \tag{3.19}$$

and

$$\frac{\partial^2 y}{\partial x_i \partial x_j} = \frac{1+\rho}{y}\frac{\partial y}{\partial x_i}\frac{\partial y}{\partial x_j}. \tag{3.20}$$

The CES production function

$$\theta y = \beta_0 \left(\sum_{i=1}^{r} \beta_i (\theta x_i)^{-\rho} \right)^{-\frac{1}{\rho}} \tag{3.21}$$

is linearly homogeneous

$$\begin{aligned} \theta y &= \beta_0 \left(\sum_{i=1}^{r} \beta_i (\theta x_i)^{-\rho} \right)^{-\frac{1}{\rho}} = \beta_0 \left(\theta^{-\rho} \sum_{i=1}^{r} \beta_i x_i^{-\rho} \right)^{-\frac{1}{\rho}} \\ &= (\theta^{-\rho})^{-\frac{1}{\rho}} \beta_0 \left(\sum_{i=1}^{r} \beta_i x_i^{-\rho} \right)^{-\frac{1}{\rho}} = \theta \beta_0 \left(\sum_{i=1}^{r} \beta_i x_i^{-\rho} \right)^{-\frac{1}{\rho}}. \end{aligned} \tag{3.22}$$

Because of being linearly homogeneous, Euler's theorem

$$\sum_{i=1}^{r} \frac{\partial y}{\partial x_i} x_i = \sum_{i=1}^{r} y^{1+\rho} \beta_0^{-\rho} \beta_i x_i^{-(1+\rho)} x_i = y \tag{3.23}$$

holds:

$$\begin{aligned} \sum_{i=1}^{r} \frac{\partial y}{\partial x_i} x_i &= \sum_{i=1}^{r} y^{1+\rho} \beta_0^{-\rho} \beta_i x_i^{-(1+\rho)} x_i = y y^{\rho} \beta_0^{-\rho} \sum_{i=1}^{r} \beta_i x_i^{-\rho} \\ &= y y^{\rho} \left[\beta_0 \left(\sum_{i=1}^{r} \beta_i x_i^{-\rho} \right)^{-\frac{1}{\rho}} \right]^{-\rho} = y y^{\rho} y^{-\rho} = y. \end{aligned} \tag{3.24}$$

The ratio of inputs x_i and x_j can be obtained from the marginal rate of technical substitution

$$\frac{\beta_i}{\beta_j}\left(\frac{x_j}{x_i}\right)^{(1+\rho)} = \frac{w_i}{w_j} \Rightarrow \frac{x_i}{x_j} = \left(\frac{\beta_i}{\beta_j}\right)^{\frac{1}{1+\rho}} \left(\frac{w_j}{w_i}\right)^{\frac{1}{1+\rho}}. \tag{3.25}$$

Through differentiation, we obtain

$$\begin{aligned}\frac{\partial(x_i/x_j)}{\partial(w_j/w_i)} &= \frac{1}{1+\rho}\left(\frac{\beta_i}{\beta_j}\right)^{\frac{1}{1+\rho}}\left(\frac{w_j}{w_i}\right)^{\frac{1}{1+\rho}-\frac{1+\rho}{1+\rho}} \\ &= \frac{1}{1+\rho}\left(\frac{\beta_i}{\beta_j}\right)^{\frac{1}{1+\rho}}\left(\frac{w_j}{w_i}\right)^{\frac{-\rho}{1+\rho}}\end{aligned} \tag{3.26}$$

and the elasticity of substitution as

$$\begin{aligned}\sigma = \frac{\frac{\partial(x_i/x_j)}{\partial(w_j/w_i)}}{\frac{x_i/x_j}{w_j/w_i}} &= \frac{\frac{1}{1+\rho}\left(\frac{\beta_i}{\beta_j}\right)^{\frac{1}{1+\rho}}\left(\frac{w_j}{w_i}\right)^{\frac{-\rho}{1+\rho}}}{\frac{\left(\frac{\beta_i}{\beta_j}\right)^{\frac{1}{1+\rho}}\left(\frac{w_j}{w_i}\right)^{\frac{1}{1+\rho}}}{\frac{w_j}{w_i}}} \\ &= \frac{\frac{1}{1+\rho}\left(\frac{\beta_i}{\beta_j}\right)^{\frac{1}{1+\rho}}\left(\frac{w_j}{w_i}\right)^{\frac{-\rho}{1+\rho}}\frac{w_j}{w_i}}{\left(\frac{\beta_i}{\beta_j}\right)^{\frac{1}{1+\rho}}\left(\frac{w_j}{w_i}\right)^{\frac{1}{1+\rho}}} = \frac{1}{1+\rho}.\end{aligned} \tag{3.27}$$

Looking again at the marginal product

$$\frac{\partial y}{\partial x_i} = y^{1+\rho}\beta_0^{-\rho}\beta_i x_i^{-(1+\rho)} = \frac{w_i}{p} \tag{3.28}$$

we find (with the definition of σ in Eq. (3.27))

$$\begin{aligned}(1+\rho)\ln y - \rho\ln\beta_0 + \ln\beta_i - (1+\rho)\ln x_i &= \ln w_i - \ln p \\ \frac{1}{\sigma}\ln y - \frac{1}{\sigma}\ln x_i &= \rho\ln\beta_0 - \ln\beta_i + \ln w_i - \ln p \\ \ln y - \ln x_i &= \sigma\rho\ln\beta_0 - \sigma\ln\beta_i + \sigma\,(\ln w_i - \ln p) \\ \ln\left(\frac{y}{x_i}\right) &= \sigma\ln\left(\frac{\beta_0^\rho}{\beta_i}\right) + \sigma\,(\ln w_i - \ln p) \\ &= -\sigma\ln\left(\frac{\beta_i}{\beta_0^\rho}\right) + \sigma\ln\left(\frac{w_i}{p}\right).\end{aligned} \tag{3.29}$$

Note that the regression coefficient of the regression of logarithmic factor productivity on logarithmic real wages is the constant elasticity of substitution which in the Cobb–Douglas is 1 by definition.

Using again the marginal product for input x_i, we obtain the factors share in output

$$\begin{aligned}\frac{\partial y}{\partial x_i} &= y^{1+\rho}\beta_0^{-\rho}\beta_i x_i^{-(1+\rho)} = \frac{w_i}{p}\\ \frac{w_i x_i}{py} &= y^{\rho}\beta_0^{-\rho}\beta_i x_i^{-\rho} = \left(\frac{\beta_i}{\beta_0^{\rho}}\right)^{\sigma}\left(\frac{w_i}{p}\right)^{\sigma-1}\end{aligned} \tag{3.30}$$

and using the marginal rate of technical substitution, we find the ratio of input shares

$$\frac{x_i w_i}{x_j w_j} = \left(\frac{\beta_i}{\beta_j}\right)^{\frac{1}{1+\rho}}\left(\frac{w_j}{w_i}\right)^{\frac{1}{1+\rho}}\left(\frac{w_j}{w_i}\right)^{-1} = \left(\frac{\beta_i}{\beta_j}\right)^{\sigma}\left(\frac{w_j}{w_i}\right)^{\sigma-1}. \tag{3.31}$$

3.3.2 Estimating the CES with R

We consider the following parametrization of the constant elasticity of substitution production function

$$y = \beta_1\left(\beta_2 X_1^{-\beta_4} + (1-\beta_2)X_2^{-\beta_4}\right)^{-\frac{\beta_3}{\beta_4}} e^u \tag{3.32a}$$

Using the log model we obtain

$$\ln y = \ln\beta_1 - \frac{\beta_3}{\beta_4}\ln\left(\beta_2 X_1^{-\beta_4} + (1-\beta_2)X_2^{-\beta_4}\right) + u. \tag{3.32b}$$

We generate data ($n = 100$) according to the CES-production function with parameter values $\beta_1 = 2.7, \beta_2 = 0.8, \beta_3 = 1.2, \beta_4 = 1$ with the following R commands:

```
n <- 100
b1 <- 2.7
b2 <- 0.8
b3 <- 1.2
b4 <- 1
set.seed(123)
x1 <- exp(runif(n))
x2 <- exp(runif(n))
u <- rnorm(n,sd=0.2)
y <- b1 * ( b2*x1^(-b4) + (1-b2)*x2^(-b4) )^(-b3/b4) *
      exp(u)
ly <- log(y)
lx1 <- log(x1)
lx2 <- log(x2)
```

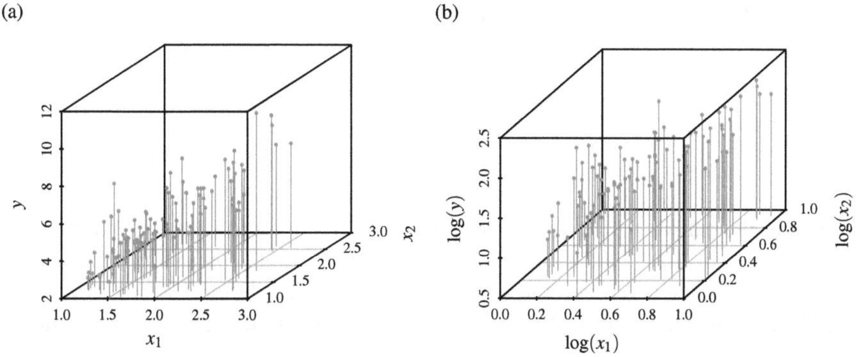

Fig. 3.2 CES production function. (**a**) Original data. (**b**) Data in logs

We display the data using a 3D-scatterplot in Fig. 3.2.

```
library(scatterplot3d)
scatterplot3d(x1, x2, y, main="",
              xlab=expression(x[1]),
              ylab=expression(x[2]),
              zlab="y",
              color="darkgrey", pch=20, type="h")
scatterplot3d(lx1, lx2, ly, main="",
              xlab=expression(paste("log(", x[1], ")")),
              ylab=expression(paste("log(", x[2], ")")),
              zlab="log(y)",
              color="darkgrey", pch=20, type="h")
```

Next we write a function `fmin` returning the sum of squared residuals (RSS) for a given parameter vector β that we want to minimize by choosing β. To avoid the occurrence of a ratio of parameters to be estimated (β_3/β_4), we estimate this ratio using a single parameter β_3 from which the original parameter can easily be obtained

$$\begin{aligned} \text{RSS} &= \sum u^2 \\ &= \sum \left(\ln y - \left[\ln \beta_1 - \beta_3 \ln\left(\beta_2 X_1^{-\beta_4} + (1-\beta_2) X_2^{-\beta_4}\right)\right]\right)^2 . \end{aligned} \tag{3.33}$$

```
fmin <- function(b){
  sum((ly-(log(b[1])-b[3]*log(b[2]*x1^(-b[4])
                              +(1-b[2])*x2^(-b[4]))))^2)
}
```

Using `optim()` and a rather robust default algorithm ("Nelder–Mead") without providing analytic derivatives, we obtain the following parameter estimates when providing the true parameters as starting values:

```
#b1 <- 2.7; b2 <- 0.8; b3 <- 1.2; b4 <- 1
bstart <- c(2.7,0.8,1.2,1)
optim(bstart, fmin)$par

## [1] 2.6243 0.8083 1.2873 0.9413
```

As an alternative to the function `optim()`, we use the function `maxLik()` implemented in the maxLik package. As we now want to maximize instead of minimizing, we have to revert the sign of our function and define `fmax` accordingly. The default algorithm is the Newton–Raphson algorithm discussed in some detail above.

```
library(maxLik)
fmax <- function(b){-sum((ly-(log(b[1])-b[3]*log(b[2]*x1^(-b[4])
                          +(1-b[2])*x2^(-b[4]))))^2)}
bstart <- c(2.7,0.8,1.2,1)
summary(maxLik(fmax, start=bstart))

## --------------------------------------------
## Maximum Likelihood estimation
## Newton--Raphson maximisation, 7 iterations
## Return code 2: successive function values within tolerance limit
## Log-Likelihood: -3.699
## 4  free parameters
## Estimates:
##      Estimate Std. error t value   Pr(> t)
## [1,]    2.624      0.585    4.49 0.0000072 ***
## [2,]    0.808      0.191    4.24 0.0000227 ***
## [3,]    1.289      7.020    0.18      0.85
## [4,]    0.940      5.057    0.19      0.85
## ---
## Signif. codes:
## 0 '***' 0.001 '**' 0.01 '*' 0.05 '.' 0.1 ' ' 1
## --------------------------------------------
```

Choosing some other starting values results in varying parameter estimates indicating a rather badly behaved function we want to minimize.

```
bstart <- c(0.5,0.5,0.5,0.5)
optim(bstart, fmin)$par

## [1] 2.5991 0.8061 2.7561 0.4415

bstart <- rep(1,4)
optim(bstart, fmin)$par

## [1] 2.6371 0.8108 1.0507 1.1495
```

While the estimated parameter values for β_1 and β_2 are almost identical with different starting values, the estimates for β_3 and β_4 differ strongly. We draw the residual sum of squares we want to minimize as a function of β_i holding the other parameter values constant at their true values[2]:

```
nn <- 101
b1v <- seq(2.2,3.2,length=nn)
b2v <- seq(0.3,1.3,length=nn)
b3v <- seq(0.7,1.7,length=nn)
b4v <- seq(0.5,1.5,length=nn)
lb1v <- length(b1v)
be <- matrix(NA,lb1v,4)
for (i in 1:lb1v){
  be[i,] <- c(fmin(c(b1v[i],0.8,1.2,1)),
    fmin(c(2.7,b2v[i],1.2,1)),
    fmin(c(2.7,0.8,b3v[i],1)),
    fmin(c(2.7,0.8,1.2,b4v[i])))
  }
```

The plot is obtained by the following code and can be seen in Fig. 3.3.

```
plot(b1v,be[,1],type="l",xlab=expression(beta[1]),
    ylab="RSS",main="")
plot(b2v,be[,2],type="l",xlab=expression(beta[2]),
    ylab="RSS",main="")
```

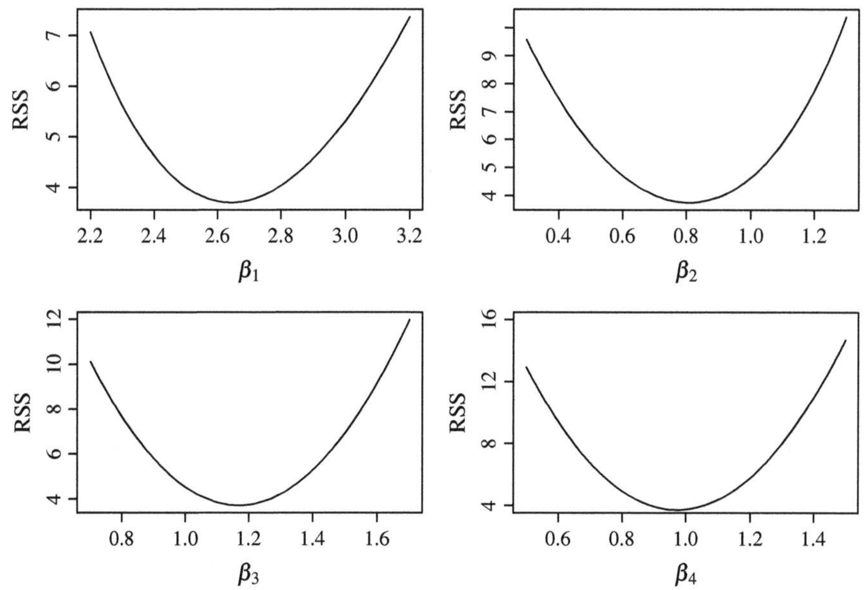

Fig. 3.3 Residual sum of squares for different parameter values

[2]In this case the residual sum of squares equals the negative partial log-likelihood if assuming normal error terms up to a scaling factor.

```
plot(b3v,be[,3],type="l",xlab=expression(beta[3]),
     ylab="RSS",main="")
plot(b4v,be[,4],type="l",xlab=expression(beta[4]),
     ylab="RSS",main="")
```

The figure of the (negative) partial log-likelihood plots reveal no hint for estimating problems. Note that partially varying the parameter value is not what happens in the numerical optimization process where all parameters are varied simultaneously. We now obtain the parameter values when minimizing the residual sum of squares by varying the starting values for one parameter systematically.

```
nn <- 51
b1v <- seq(2.2,3.2,length=nn)
b2v <- seq(0.3,1.3,length=nn)
b3v <- seq(0.7,1.7,length=nn)
b4v <- seq(0.5,1.5,length=nn)
lb1v <- length(b1v)
be <- matrix(NA,lb1v,4)
for (i in 1:lb1v){
  be[i,] <- c(optim(c(b1v[i],0.8,1.2,1), fmin)$par[1],
    optim(c(2.7,b2v[i],1.2,1), fmin)$par[2],
    optim(c(2.7,0.8,b3v[i],1), fmin)$par[3],
    optim(c(2.7,0.8,1.2,b4v[i]), fmin)$par[4])
  }
```

The plot is obtained by the following code and can be seen in Fig. 3.4.

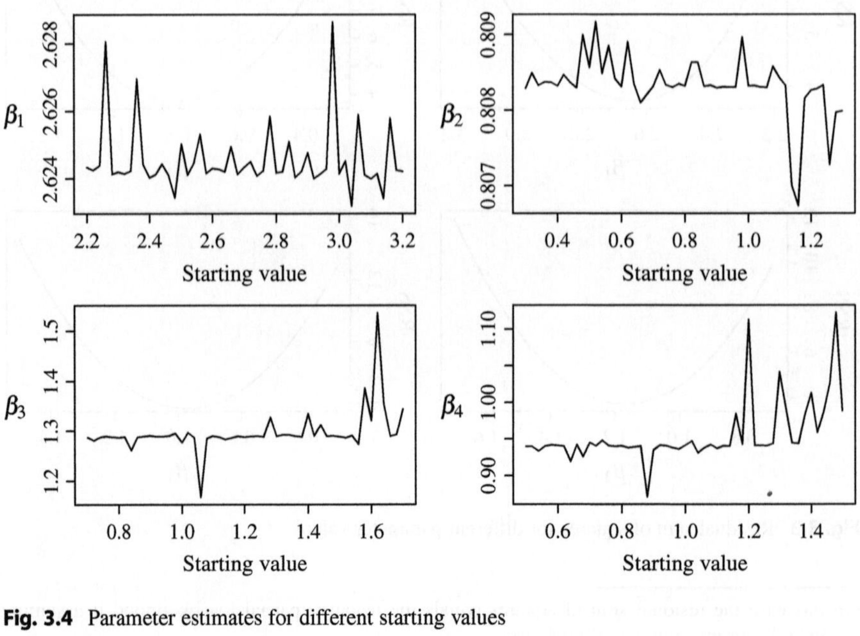

Fig. 3.4 Parameter estimates for different starting values

```
plot(b1v,be[,1],type="l",xlab="Starting value",
     ylab=expression(beta[1]),main="")
plot(b2v,be[,2],type="l",xlab="Starting value",
     ylab=expression(beta[2]),main="")
plot(b3v,be[,3],type="l",xlab="Starting value",
     ylab=expression(beta[3]),main="")
plot(b4v,be[,4],type="l",xlab="Starting value",
     ylab=expression(beta[4]),main="")
```

Obviously, the numerical optimization procedure for obtaining the parameter estimates of the CES production function is highly fragile. This may be one of the reasons rather few empirical applications of the CES production function can be found in the literature. Finally, we look at the correlation matrix of the parameter estimates which is obtained using a numerical approximation in the `optim()` routine.

```
bstart <- c(2.7,0.8,1.2,1)
V <- optim(bstart, fmin, hessian=T)$hessian
round(cov2cor(V),4)

##        [,1]   [,2]   [,3]   [,4]
## [1,] 1.0000 0.0765 0.9042 0.8998
## [2,] 0.0765 1.0000 0.3214 0.3214
## [3,] 0.9042 0.3214 1.0000 0.9997
## [4,] 0.8998 0.3214 0.9997 1.0000
```

We find that several pairs of parameters are highly correlated (some are almost linear functions) and therefore cannot be identified sufficiently in the numerical optimization process.

3.4 The Transcendental Logarithmic Production Function

So far, we have considered the Cobb–Douglas and the CES production function. Both functions are restrictive because of the following assumptions:

- Homotheticity: factor shares of output are independent of total output
- Additivity (Separability): the elasticities of substitution are constant and equal for any pair of inputs

Typical specifications of aggregate value added functions assume these restrictions (homotheticity, additivity) to hold. Berndt and Christensen (1973) and Christensen et al (1973) suggest a translog second order approximation of quantities and price functions. In this approximate production function homotheticity and additivity are not imposed as restrictions generally but can be imposed and tested.

Using duality, it can be shown that the production and the price function are identical with respect to shares of inputs and partial elasticities of substitution. In the literature, the translog function is regarded as superior compared to Cobb–Douglas and CES production functions because of its flexibility. When estimating the translog function, one usually faces the problem of a large number of regression coefficients and highly correlated regressors.

3.4.1 Properties of the Function

The function is given as

$$\begin{aligned}\ln y &= \beta_0 + \sum_{i=1}^{r} \beta_i \ln x_i + \frac{1}{2}\sum_{i=1}^{r} \beta_{ii} (\ln x_i)^2 + \frac{1}{2}\sum_{i=1}^{r}\sum_{j=1,i\neq j}^{r} \beta_{ij} \ln x_i \ln x_j \\ &= \beta_0 + \sum_{i=1}^{r} \beta_i \ln x_i + \frac{1}{2}\sum_{i=1}^{r} \beta_{ii} (\ln x_i)^2 + \sum_{i=1}^{r}\sum_{j=1,i<j}^{r} \beta_{ij} \ln x_i \ln x_j \\ &= \beta_0 + \sum_{i=1}^{r} \beta_i \ln x_i + \frac{1}{2}\sum_{i=1}^{r}\sum_{j=1}^{r} \beta_{ij} \ln x_i \ln x_j\end{aligned} \tag{3.34}$$

with $\beta_{ij} = \beta_{ji}$. The derivative of the logarithm of the production with respect to input i is

$$\frac{\partial \ln y}{\partial x_i} = \frac{1}{x_i}\beta_i + \frac{1}{x_i}\sum_{j=1}^{r} \beta_{ij} \ln x_j. \tag{3.35}$$

Using the following relation

$$\frac{\partial \ln y}{\partial y} = \frac{1}{y} \Rightarrow \partial \ln y = \frac{\partial y}{y} \Rightarrow y\partial \ln y = \partial y \tag{3.36}$$

we obtain the following condition

$$\frac{\partial y}{\partial x_i} = \frac{y}{x_i}\left(\beta_i + \sum_{j=1}^{r} \beta_{ij} \ln x_j\right) > 0. \tag{3.37}$$

For the marginal product of input i to be positive, this condition must hold. Only the term in parenthesis must be verified to be positive because $y > 0$ and $x_i > 0$.

Assuming that factor prices equal their marginal products, the term in brackets is factor i's share (S_i) in output

$$\begin{aligned}\frac{\partial y}{\partial x_i} &= \frac{w_i}{p} = \frac{y}{x_i}\left(\beta_i + \sum_{j=1}^{r} \beta_{ij} \ln x_j\right) > 0 \\ S_i &= \frac{x_i w_i}{yp} = \beta_i + \sum_{j=1}^{r} \beta_{ij} \ln x_j > 0.\end{aligned} \tag{3.38}$$

Therefore, the assumption of monotonicity can be tested empirically. The second order derivatives are

$$
\begin{aligned}
\frac{\partial^2 y}{\partial x_i \partial x_i} &= \frac{\partial \frac{y}{x_i}\left(\beta_i + \sum_{j=1}^{r} \beta_{ij} \ln x_j\right)}{\partial x_i} \\
&= \left(\frac{\frac{\partial y}{\partial x_i} x_i - y}{x_i^2}\right)\left(\beta_i + \sum_{j=1}^{r} \beta_{ij} \ln x_j\right) + \frac{y}{x_i^2}\beta_{ii} \\
&= \frac{1}{x_i^2}\,(yS_i - y)\,S_i + \frac{y}{x_i^2}\beta_{ii} = \frac{y}{x_i^2}\left((S_i - 1)S_i + \beta_{ii}\right)
\end{aligned}
\tag{3.39}
$$

and

$$
\begin{aligned}
\frac{\partial^2 y}{\partial x_i \partial x_j} &= \frac{\partial \frac{y}{x_i}\left(\beta_i + \sum_{j=1}^{r} \beta_{ij} \ln x_j\right)}{\partial x_j} \\
&= \frac{1}{x_i}\frac{\partial y}{\partial x_j}\left(\beta_i + \sum_{j=1}^{r} \beta_{ij} \ln x_j\right) + \frac{y}{x_i x_j}\beta_{ij} \\
&= \frac{1}{x_i}\frac{y}{x_j} S_j S_i + \frac{y}{x_i x_j}\beta_{ij} = \frac{y}{x_i x_j}\left(S_i S_j + \beta_{ij}\right).
\end{aligned}
\tag{3.40}
$$

The translog function is homogeneous of degree one if the following equalities are fulfilled:

$$
\begin{aligned}
& f(\theta x_1, \theta x_2, \ldots, \theta x_r) = \theta y \\
&= \ln f(\theta x_1, \theta x_2, \ldots, \theta x_r) = \ln \theta + \ln y \\
&= \beta_0 + \sum_{i=1}^{r} \beta_i \ln(\theta x_i) + \frac{1}{2}\sum_{i=1}^{r}\sum_{j=1}^{r} \beta_{ij} \ln(\theta x_i) \ln(\theta x_j) \\
&= \beta_0 + \sum_{i=1}^{r} \beta_i \left(\ln \theta + \ln x_i\right) + \frac{1}{2}\sum_{i=1}^{r}\sum_{j=1}^{r} \beta_{ij} \left(\ln \theta + \ln x_i\right)\left(\ln \theta + \ln x_j\right) \\
&= \beta_0 + \sum_{i=1}^{r} \beta_i \ln x_i + \ln \theta \sum_{i=1}^{r} \beta_i \\
&\quad + \frac{1}{2}\sum_{i=1}^{r}\sum_{j=1}^{r} \beta_{ij}\left[(\ln \theta)^2 + \ln \theta \ln x_i + \ln \theta \ln x_j + \ln x_i \ln x_j\right]
\end{aligned}
\tag{3.41}
$$

$$
\begin{aligned}
&= \beta_0 + \sum_{i=1}^{r} \beta_i \ln x_i + \frac{1}{2} \sum_{i=1}^{r} \sum_{j=1}^{r} \beta_{ij} \ln x_i \ln x_j \\
&+ \ln \theta \sum_{i=1}^{r} \beta_i + (\ln \theta)^2 \frac{1}{2} \sum_{i=1}^{r} \sum_{j=1}^{r} \beta_{ij} \\
&+ \ln \theta \frac{1}{2} \sum_{i=1}^{r} \ln x_i \sum_{j=1}^{r} \beta_{ij} + \ln \theta \frac{1}{2} \sum_{j=1}^{r} \ln x_j \sum_{i=1}^{r} \beta_{ij}.
\end{aligned}
$$

Therefore, the restrictions are

$$
\begin{aligned}
&\sum_{i=1}^{r} \beta_i = 1, \quad \sum_{i=1}^{r} \sum_{j=1}^{r} \beta_{ij} = 0, \quad \sum_{j=1}^{r} \beta_{ij} = 0 \text{ for } i = 1, \ldots, n \quad \text{and} \\
&\sum_{i=1}^{r} \beta_{ij} = 0 \text{ for } j = 1, \ldots, n.
\end{aligned}
\tag{3.42}
$$

The function is separable if the marginal rate of technical substitution between any two inputs is independent of the amount used of the other inputs. We consider the case with three inputs x_1, x_2, and x_3. The condition for weak separability is

$$
\frac{\partial y / \partial x_1}{\partial y / \partial x_2} = \frac{\frac{y}{x_1}\left(\beta_1 + \sum_{j=1}^{r} \beta_{1j} \ln x_j\right)}{\frac{y}{x_2}\left(\beta_2 + \sum_{j=1}^{r} \beta_{2j} \ln x_j\right)} = \frac{\beta_1 x_2 + x_2 \sum_{j=1}^{r} \beta_{1j} \ln x_j}{\beta_2 x_1 + x_1 \sum_{j=1}^{r} \beta_{2j} \ln x_j} \tag{3.43}
$$

and

$$
\begin{aligned}
0 &= \partial \frac{\frac{\partial y / \partial x_1}{\partial y / \partial x_2}}{\partial x_3} = \partial \frac{\frac{\beta_1 x_2 + x_2 \sum_{j=1}^{r} \beta_{1j} \ln x_j}{\beta_2 x_1 + x_1 \sum_{j=1}^{r} \beta_{2j} \ln x_j}}{\partial x_3} \\
&= \frac{x_2 \beta_{13} \frac{1}{x_3} x_1 S_2 - x_1 \beta_{23} \frac{1}{x_3} x_2 S_1}{x_1^2 S_2^2} = \frac{\frac{x_2}{x_3}\left(\beta_{13} S_2 - \beta_{23} S_1\right)}{x_1 S_2^2}.
\end{aligned}
\tag{3.44}
$$

Assuming that all inputs and S_2 are positive, the condition simplifies towards

$$
\beta_{13} S_2 - \beta_{23} S_1 = 0. \tag{3.45}
$$

Hence, the condition is met if $\beta_{13} = \beta_{23} = 0$. If considering the MRTS between x_1 and x_3 and between x_2 and x_3, we obtain the conditions

$$
\beta_{12} = \beta_{13} = \beta_{23} = 0. \tag{3.46}
$$

Now using the definition of income shares, the condition is also met if $\beta_{13} \neq 0, \beta_{23} \neq 0$ but

$$\begin{aligned} 0 &= \beta_{13}\left(\beta_2 + \sum_{j=1}^{r} \beta_{2j} \ln x_j\right) - \beta_{23}\left(\beta_1 + \sum_{j=1}^{r} \beta_{1j} \ln x_j\right) \\ &= \beta_{13}\left(\beta_2 + \beta_{21}\ln x_1 + \beta_{22}\ln x_2 + \beta_{23}\ln x_3\right) \\ &\quad - \beta_{23}\left(\beta_1 + \beta_{11}\ln x_1 + \beta_{12}\ln x_2 + \beta_{13}\ln x_3\right) \\ &= \beta_2\beta_{13} - \beta_1\beta_{23} + (\beta_{13}\beta_{21} - \beta_{23}\beta_{11})\ln x_1 \\ &\quad + (\beta_{13}\beta_{22} - \beta_{23}\beta_{12})\ln x_2 + (\beta_{13}\beta_{23} - \beta_{23}\beta_{13})\ln x_3 \end{aligned} \tag{3.47}$$

and

$$\begin{aligned} \beta_2\beta_{13} - \beta_1\beta_{23} &= 0 \\ \beta_{13}\beta_{21} - \beta_{23}\beta_{11} &= 0 \\ \beta_{13}\beta_{22} - \beta_{23}\beta_{12} &= 0. \end{aligned} \tag{3.48}$$

Manipulation of the conditions results in

$$\begin{aligned} &\frac{\beta_1}{\beta_2} = \frac{\beta_{11}}{\beta_{12}} = \frac{\beta_{12}}{\beta_{22}} = \frac{\beta_{13}}{\beta_{23}} \\ &\beta_{12} = \beta_{11}\frac{\beta_2}{\beta_1}; \beta_{22} = \beta_{11}\frac{\beta_2^2}{\beta_1^2}; \beta_{13} = \frac{\beta_1}{\beta_2}\beta_{23}. \end{aligned} \tag{3.49}$$

This condition is referred to as the nonlinear separability condition whereas $\beta_{13} = \beta_{23} = 0$ is referred to as the linear separability condition. Note that if these conditions are not met, the estimation of substitution parameters between inputs ignoring further relevant inputs will result in misleading parameter estimates.

3.4.2 Estimating the TL with R

The transcendental logarithmic function with two inputs x_1 and x_2 is

$$\begin{aligned} \ln y = \beta_0 &+ \beta_1 \ln x_1 + \beta_2 \ln x_2 + \beta_3 (\ln x_1)^2 \\ &+ \beta_4 (\ln x_2)^2 + \beta_5 \ln x_1 \ln x_2. \end{aligned} \tag{3.50}$$

To estimate the function, we use the data generated according to the CES-production function.

```
n <- 100
b0 <- log(2.7)
b1 <- 0.8
b2 <- 1.2
b3 <- 1
set.seed(123)
x1 <- exp(runif(n))
x2 <- exp(runif(n))
u <- rnorm(n,sd=0.2)
y <- exp(b0) *
      ( b1*x1^(-b4) + (1-b1)*x2^(-b3) )^(-b2/b3) *
      exp(u)
ly <- log(y)
```

Because the TL is linear in parameters it can be estimated by ordinary least squares using the function `lm()`. Before estimating, we calculate the transformed variables necessary for the translog production function:

```
lx1 <- log(x1)
lx2 <- log(x2)
lx1_2 <- lx1^2
lx2_2 <- lx2^2
lx1lx2 <- lx1*lx2
reg1 <- lm(ly~lx1+lx2+lx1_2+lx2_2+lx1lx2)
summary(reg1)

##
## Call:
## lm(formula = ly ~ lx1 + lx2 + lx1_2 + lx2_2 + lx1lx2)
##
## Residuals:
##      Min       1Q   Median       3Q      Max
## -0.3893 -0.1376 -0.0332  0.1149  0.6702
##
## Coefficients:
##               Estimate Std. Error t value    Pr(>|t|)
## (Intercept)   0.84458    0.14555    5.80 0.000000088
## lx1           1.11269    0.36428    3.05      0.0029
## lx2           0.62521    0.36553    1.71      0.0905
## lx1_2        -0.12382    0.29025   -0.43      0.6706
## lx2_2        -0.36232    0.29743   -1.22      0.2262
## lx1lx2       -0.00308    0.28609   -0.01      0.9914
##
## (Intercept) ***
## lx1         **
## lx2         .
## lx1_2
## lx2_2
## lx1lx2
## ---
## Signif. codes:
## 0 '***' 0.001 '**' 0.01 '*' 0.05 '.' 0.1 ' ' 1
##
```

```
## Residual standard error: 0.197 on 94 degrees of freedom
## Multiple R-squared:  0.684,      Adjusted R-squared:  0.667
## F-statistic: 40.7 on 5 and 94 DF,  p-value: <2e-16
```

We know that the data have not been generated according to the Cobb–Douglas production function. Nevertheless, we test this hypothesis. In the Cobb–Douglas case the parameter restriction $\beta_3 = \beta_4 = \beta_5 = 0$ holds. In this case, a simple F-test can be used to test whether a restricted model has a significantly higher sum of squared residuals. Denoting the vector of estimated residuals from the unrestricted model by $\hat{\mathbf{u}}$ and from the restricted model by $\hat{\mathbf{u}}_*$, the test statistic is

$$F_{k-k_*,n-k} = \frac{(\hat{\mathbf{u}}_*'\hat{\mathbf{u}}_* - \hat{\mathbf{u}}'\hat{\mathbf{u}})/(k-k_*)}{(\hat{\mathbf{u}}'\hat{\mathbf{u}})/(n-k)}. \tag{3.51}$$

$k - k_*$ is the number of restrictions (here 3) and k is the number of regression parameters in the unrestricted regression (here 6). In R, we use the following commands

```
reg2 <- lm(ly~lx1+lx2)
u <- matrix(reg1$resid)
us <- matrix(reg2$resid)
k <- length(coef(reg1))
ks <- length(coef(reg2))
F.stat <- as.vector(((t(us)%*%us-t(u)%*%u)/(k-ks))/
                    (t(u)%*%u/(n-k)))
F.stat

## [1] 0.5225

pvalue <- 1-pf(F.stat,k-ks,n-k)
pvalue

## [1] 0.6679
```

The result indicates that one would obtain a difference in sums of squares at least as large as we have obtained with our data with a probability of about 67 % if the data had been generated with the Cobb–Douglas model. A much nicer way to obtain the test result is using the function `anova()`:

```
anova(reg2,reg1)

## Analysis of Variance Table
##
## Model 1: ly ~ lx1 + lx2
## Model 2: ly ~ lx1 + lx2 + lx1_2 + lx2_2 + lx1lx2
##   Res.Df  RSS Df Sum of Sq    F Pr(>F)
## 1     97 3.72
## 2     94 3.66  3     0.061 0.52   0.67
```

The translog function with restriction $\beta_3 = \beta_4 = -0.5\beta_5$ has been shown by Kmenta to be a reasonable approximation to the CES-production function. This can be tested in R using the methodology of testing linear restrictions.

```
b <- matrix(coef(reg1))
v <- vcov(reg1)
R <- rbind(c(0,0,0,1,-1,0),c(0,0,0,0,1,0.5))
r <- matrix(c(0,0))
q <- nrow(R)
n_k <- length(y)-length(coef(reg1))
Fs <- as.vector(1/q*t(R%*%b-r)%*%
                  solve(R%*%v%*%t(R))%*%(R%*%b-r))

Fs

## [1] 0.5879

1-pf(Fs,q,n_k)

## [1] 0.5575
```

The number of parameters increases fast with an increasing number of inputs. Including the intercept, the number of parameters with k inputs is

$$\frac{k(k+3)}{2} + 1. \tag{3.52}$$

Transforming the inputs into regressor variables is tedious with $k > 2$. In this case, one might want to use the function `translogEst()` available in the package micEcon. The function sets up the model matrix automatically when given the k input variables. Note that the squared regressors are multiplied by factor 0.5. The following R-code results in identical estimated parameters (using again the artificial CES-data):

```
library(micEcon)
dat <- data.frame(y,x1,x2)
TLE <- translogEst(yName="y",xNames=c("x1","x2"),
                   data=dat)
LM <- lm(ly~lx1+lx2+I(0.5*lx1_2)+lx1lx2+I(0.5*lx2_2))
e <- data.frame(TLE=coef(TLE),LM=coef(LM))
e

##              TLE        LM
## a_0     0.844575  0.844575
## a_1     1.112692  1.112692
## a_2     0.625206  0.625206
## b_1_1 -0.247641 -0.247641
## b_1_2 -0.003078 -0.003078
## b_2_2 -0.724631 -0.724631
```

3.5 Recommended Readings

The seminal papers for the discussed production functions are for the Cobb–Douglas production function

- Cobb CW, Douglas PH (1928) A theory of production. Am Econ Rev 18:139–165,

for the constant elasticity of substitution function

- Arrow KJ, Chenery HB, Minhas BS, Solow RM (1961) Capital–labor substitution and economic efficiency. Rev Econ Stat 43(3):225–250,

and for the translog production function

- Christensen LR, Jorgensen DW, Lau JL (1971) Conjugate duality and the transcendental logarithmic production function. Econometrica 39:255–256

and

- Christensen LR, Jorgensen DW, L LJ (1973) Transcendental logarithmic production frontiers. Rev Econ Stat 55(1):28–45.

The standard textbook on theoretical production analysis is

- Chambers R (1988) Applied production analysis. A dual approach. Press Syndicate of the University of Cambridge, Cambridge.

A concise discussion of the most important utility and production functions and their characteristics is provided by

- Chung JW (1994) Utility and production functions. Blackwell, Oxford.

A very intuitive presentation of regression analysis with R is provided by

- Fox J, Weisberg S (2011) An R companion to applied regression. Sage Publications, Thousand Oaks.

Nonlinear regression models are discussed by

- Ritz C, Streibig JC (2008) Nonlinear regression with R. Springer, New York.

3.6 Exercises

In this exercise, we will use a single cross-section (1997) of the data set "riceProdPhil" available in the package frontier. The variables we use are:

- PROD output (tonnes of freshly threshed rice)
- AREA area planted (hectares)
- LABOR labor used (man-days of family and hired labor)

1. The data structure
   a. Load the data from year 1997 into object d using the following command:

```
install.packages("frontier")
data("riceProdPhil", package = "frontier")
d <- subset(riceProdPhil,YEARDUM==8,
            select=c(PROD,LABOR,AREA))
```

   b. Try to get an overview of the data through
      i. using `summary()`
      ii. studying the distributions of the three variables LABOR (x_1), AREA (x_2), PROD (y) (hint: use `hist()` or `density()`).
      iii. studying the distributions of the three variables in logarithms.
   c. Draw bivariate plots of (x_1, y) and (x_2, y) using
      i. original and
      ii. logarithmic values.
   d. Calculate the variance–covariance matrix and the correlation matrix of x_1, x_2, y in
      i. original and
      ii. logarithmic values.
2. Consider a simple two-factor Cobb–Douglas production function

$$y = e^{\beta_0} x_1^{\beta_1} x_2^{\beta_2} e^u.$$

   a. Estimate the regression function using `lm()` and discuss the detailed results contained in the summary of the regression.
   b. Plot the distribution of the estimated residuals.
   c. Estimate the production function under the restriction $\beta_1 + \beta_2 = 1$ and discuss the results.
   d. Display the residual distribution and the bivariate plot of regressors graphically.
   e. Re-fit the restricted regression after dropping the outlier. Are there any eye-catching differences in results?
3. Consider the constant elasticity of substitution production function

$$y = \beta_0 \left(\beta_1 x_1^{-\beta_3} + (1 - \beta_1) x_2^{-\beta_3} \right)^{-\frac{\beta_2}{\beta_3}} e^u.$$

a. Obtain the regression function after taking the logarithm.
b. Try to obtain the ML-estimates of the parameters using the function `optim()`. Hints: (1) try several different starting values. (2) As constants do not change during the numerical optimization, it is sufficient to minimize $\sum(y - f(x))^2$ for maximizing the log-likelihood. (3) To prevent ratios of parameters to be estimated, use $-b_2$ instead of $-b_2/b_3$ in your log-likelihood.
c. Compare the results with the ones you obtain when using `maxLik()` from the maxLik-package.
d. Plot the partial log-Likelihood for each parameter keeping the other parameter values constant at their ML-estimates.
e. How do you value your chances when trying to estimate a CES function with empirical data?

4. The translog production function.
   a. Estimate the translog production function with two inputs (x_1, x_2) using *lm*().
   b. Test whether the data could have been generated using a simple Cobb–Douglas production function.
   c. Test whether the Kmenta-approximation to the CES production function holds.

References

Arrow KJ, Chenery HB, Minhas BS, Solow RM (1961) Capital–labor substitution and economic efficiency. Rev Econ Stat 43(3):225–250

Berndt ER, Christensen LR (1973) The translog function and the substitution of equipment, structures, and labor in u.s. manufacturing 1929–68. J Econ 1:81–114

Chambers R (1988) Applied production analysis. A dual approach. Press Syndicate of the University of Cambridge, Cambridge

Christensen LR, Jorgensen DW, Lau JL (1971) Conjugate duality and the transcendental logarithmic production function. Econometrica 39:255–256

Christensen LR, Jorgensen DW, L LJ (1973) Transcendental logarithmic production frontiers. Rev Econ Stat 55(1):28–45

Chung JW (1994) Utility and production functions. Blackwell, Oxford

Cobb CW, Douglas PH (1928) A theory of production. Am Econ Rev 18:139–165

Levhari D, Samuelson PA (1966) The non-switching theorem is false. Q J Econ 80(4):518–519

Mishra SK (2007) A brief history of production functions, mPRA, North Eastern Hill Universtiy, Shillong (India) Working Paper No. 5254

Ritz C, Streibig JC (2008) Nonlinear regression with R. Springer, New York

Robinson j (1953) The production function and the theory of capital. Rev Econ Stud 21(2):81–106

Shaikh A (1974) Laws of production and laws of algebra: The humbug production function. Rev Econ Stat 56:115–120

von Thünen JH (1850) Der isolierte Staat in Beziehung auf Landwirtschaft und Natioanalökonomie. Zweiter Teil. Rostock

von Thünen JH (1910) Der isolierte Staat in Beziehung auf Landwirtschaft und Natioanalökonomie. Erster Teil: Untersuchungen über den Einfluss, den die Getreidepreise, der Reichtum des Bodens und die Abgaben auf den Ackerbau ausüben. Verlag von Gustav Fischer

Wicksell K (1923) Realkapital och kapitalränta. Ekonomisk Tisdkrift 25:145–180

Wicksell K (1934) Lectures on political economy, vol 1. G. Routledge and Sons, London
Wicksteed PH (1894) An essay on the coordination of the laws of distribution, 1932nd edn. London School of Economics and Political Science, London
Fox J, Weisberg S (2011) An R companion to applied regression. Sage Publications, Thousand Oaks

Production Functions with Panel Data

4

Abstract

In this chapter, we discuss the estimation of production functions using panel data. Panel data sets include data for producing units which have been observed for several periods. Information about differences of production relations between units and of differences between time periods allows to drop some restrictive assumptions on parameter homogeneity which have to be employed in the cases of cross sectional or time series data. We discuss the pooled, the between, the within, and the random effects estimator for static models and hint for some problems when instrumenting in dynamic panel data models. The generation of data according to the different models and their estimation are demonstrated.

4.1 Introduction

Panel data provide information about changes of inputs and output in time on the individual firm level. This additional information allows to take into account firm specific effects. Throughout this chapter, like we did in the previous chapter, we assume that firms operate technically efficient and that observed deviations from the functional production relation solely reflect pure random variations in output. Throughout this chapter, we restrict the discussion towards static models of production. In recent years, it has become rather popular to include the output of the previous period as an additional regressor in regression functions. While lagged output is almost always found to be "statistically significant", the economic meaning of lagged output is not clear.[1]

[1] An overview of dynamic panel data estimators provides Behr (2003).

A. Behr, *Production and Efficiency Analysis with R*,
DOI 10.1007/978-3-319-20502-1_4

4.2 Introduction to Panel Data

We use two indices i and t for each observation. $i\,(i = 1, \ldots, n)$ refers to a firm whereof we observe n. The index t refers to the time period $(t = 1, \ldots, T_i)$, most often years. To ease the exposition, we assume a balanced panel data set, i.e., $T_i = T$ for all i.

4.2.1 Unrestricted and Restricted Models

We assume that unit i $(i = 1, \ldots, n)$ uses the amount x_{it} of labor as input and produces output y_{it} in period t. We further assume that we have observed each firm in T different production periods $(t = 1, \ldots, T)$ and therefore have information on T different input and output quantities for this unit. To make the exposition as simple as possible, we assume a linear relationship between input and output of the following form

$$y_{it} = \beta_{0i} + x'_{it}\beta_{1i} + v_{it}. \tag{4.1}$$

In this notation, β_{0i} denotes the intercept for firm i, β_{1i} the slope parameter of firm i's production function, and v_{it} is by assumption a random error term, capturing nonsystematic deviations from the linear production function with expectation $E(v_{it} = 0)$. Using ordinary least squares (OLS), we obtain estimates of the regression parameters $\hat{\beta}_{0i}$ and $\hat{\beta}_{1i}$.

Assuming $n = 3$ and $T = 3$ for the moment, i.e., $i = 1, 2, 3$ and $t = 1, 2, 3$, we obtain three regression equations based on time series of length $T = 3$

$$y_{it} = \beta_{0i} + x'_{it}\beta_{1i} + v_{it}, \tag{4.2}$$

with three different intercepts β_{0i} and three different slope parameters β_{1i}. Note that we assume the regression parameters β_{0i} and β_{1i} to be the same in all time periods t when estimating the unit specific regressions based on time series data.

We can also look at the cross-sectional regression equations. For each period t, we obtain one cross-sectional regression

$$y_{it} = \beta_{0t} + x'_{it}\beta_{1t} + v_{it}. \tag{4.3}$$

If we use cross-sectional data in a single period t, we assume the parameters β_{0t} and β_{1t} to be the same for all different units i.

Combining the data for all T periods and all n units, we obtain a data set with $n \times T$ observations. Note that the general model

$$y_{it} = \beta_{0it} + x'_{it}\beta_{1it} + v_{it} \tag{4.4}$$

cannot be estimated, as there would be $n \times T$ intercepts and $n \times T$ slope parameters to be estimated with only $n \times T$ observations available. Therefore, some restrictions have to be imposed on the model.

4.2.1.1 Introducing Error Components

So far, we have only used v_{it} to denote the error term of the regression equation. In the panel data literature it is common to use a notation allowing to talk about possible components of the error term v_{it}. The notation $v_{it} = \alpha_i + \lambda_t + u_{it}$ introduces a unit specific term α_i and a period specific term λ_t. The u_{it} is assumed to be unit and period specific. Using β_0 for an overall intercept and assuming unit specific slope parameters for the moment, the model can be written as

$$y_{it} = \beta_0 + x_{it}'\beta_{1i} + \alpha_i + \lambda_t + u_{it}. \tag{4.5}$$

Based on this equation, one can make different assumptions on the error terms. For example, assuming $\lambda_t = 0$ for all $t = 1, \ldots, T$ and including α_i leads to a model with unit specific intercepts

$$y_{it} = (\beta_0 + \alpha_i) + x_{it}'\beta_{1i} + u_{it}. \tag{4.6}$$

Assuming $\alpha_i = 0$ for all $i = 1, \ldots, n$ and regarding λ_t leads to a model with time specific intercepts

$$y_{it} = (\beta_0 + \lambda_t) + x_{it}'\beta_{1i} + u_{it}. \tag{4.7}$$

If α_i and λ_t are included, the intercepts will be unit and period specific

$$y_{it} = (\beta_0 + \alpha_i + \lambda_t) + x_{it}'\beta_{1i} + u_{it}. \tag{4.8}$$

4.2.2 Pooled, Within, Between

The panel data set provides information on n firms observed in T periods. This allows to estimate different models. We could ignore that fact that firms and periods may be specific and pool all the data. This results in the pooled model. We could regard the firms as specific and assume that each firm has an individual intercept but that all firms have identical slope parameters. This is what is assumed in the within model. We could also use the T observations for each firm to calculate mean values of inputs and output for each firm across time and use the n observations (averages) to estimate a production function. This is called the between model.

4.2.2.1 Pooled

The most restrictive model for panel data is the pooled model that restricts the intercept and the slope parameter to be identical for all units i and all periods t

$$y_{it} = \beta_0 + x_{it}'\beta_1 + v_{it}. \tag{4.9}$$

This model can be estimated by OLS. Note that if there had been error components

$$v_{it} = \alpha_i + \lambda_t + u_{it} \tag{4.10}$$

the residuals for a single unit i would be correlated because the same unit specific component α_i is contained in each of the T observations for firm i. For a single time period t, the residuals would be correlated because of the common error component λ_t contained in each of the n observations for this period.

4.2.2.2 Within

The within estimator only uses the variation within units and eliminates the differences in levels between units. The within estimation is also called the fixed effects estimator. In this case the slope parameter is assumed to be identical for all units and all periods, but the intercept is allowed to vary across the n units

$$y_{it} = \beta_{0i} + x'_{it}\beta_1 + u_{it}. \tag{4.11}$$

Using the notation for the error components model, here $v_{it} = \alpha_i + u_{it}$ and assuming $\lambda_t = 0$, we can express the unit specific intercept β_{0i} as the sum of the overall intercept β_0 and the fixed unit specific error component α_i

$$y_{it} = (\beta_0 + \alpha_i) + x'_{it}\beta_1 + u_{it}. \tag{4.12}$$

Note that with n units we can either include an overall intercept and $n - 1$ fixed components α_i or n fixed components omitting an overall intercept. Both model matrices will result in n different intercepts for the n units. For estimators to have nice properties, we most often want to include an overall intercept.

The model allows to capture for the n units effects which are constant across time, but vary over units. In the context of production functions, one could, e.g., think about different qualities of soil that lead to constant differences in logarithmic output between units at all levels of production. Note that the slope parameter can be obtained either through the inclusion of dummy variables (variables taking only the values 0 or 1) for units or equivalently through demeaning the data before estimation. When demeaning the data, we subtract for each individual firm i its specific mean across time from the observations.

4.2.2.3 Between

The between estimator only uses the variation across units and eliminates the variation within units across time. The average level for units is captured by the unit specific means across time

$$\bar{y}_i = \beta_0 + \bar{x}'_t\beta_1 + u_i. \tag{4.13}$$

4.2.3 Random Effects

We can generate data according to a specific model using random number generators. With random number generators, we can, e.g., generate n realizations of a random variate following a previously specified distribution (e.g., the normal distribution). Often economists regard empirical data as if they have been generated using random number generators.

4.2.3.1 Random Error Components

We have already introduced the notation for unit and time components

$$v_{it} = \alpha_i + \lambda_t + u_{it}. \tag{4.14}$$

In fixed effects models, the unit effects α_i and time effects λ_t are regarded as fixed for units/periods carrying information specific to units/periods. In random effects models, these effects are regarded as the result of random processes (generation of random numbers by the use of a random number generator) with the following properties:

$$\begin{aligned}
\mathrm{E}(\alpha_i) &= \mathrm{E}(\lambda_t) = \mathrm{E}(u_{it}) = 0 \\
\mathrm{E}(\alpha_i\lambda_t) &= \mathrm{E}(\alpha_i u_{it}) = \mathrm{E}(\lambda_t u_{it}) = 0 \\
\mathrm{E}\left(\alpha_i\alpha_j\right) &= \begin{cases} \sigma_\alpha^2, & \text{if } i = j \\ 0, & \text{if } i \neq j \end{cases} \\
\mathrm{E}(\lambda_t\lambda_s) &= \begin{cases} \sigma_\lambda^2, & \text{if } t = s \\ 0, & \text{if } t \neq s \end{cases} \\
\mathrm{E}\left(u_{it}u_{js}\right) &= \begin{cases} \sigma_u^2, & \text{if } i = j, t = s \\ 0, & \text{otherwise} \end{cases} \\
\mathrm{E}\left(\alpha_i x_{it}'\right) &= \mathrm{E}\left(\lambda_t x_{it}'\right) = \mathrm{E}\left(u_{it}x_{it}'\right) = \mathbf{0}'
\end{aligned} \tag{4.15}$$

Based on these assumptions, the total variance of the composed error term v_{it} is the sum of the variances of unit effects α_i, time effects λ_t and idiosyncratic error u_{it}

$$\begin{aligned}
\mathrm{E}\left(v_{it}^2\right) &= \mathrm{E}\left(\left[\alpha_i + \lambda_t + u_{it}\right]^2\right) \\
&= \mathrm{E}\left(\alpha_i^2 + \lambda_t^2 + u_{it}^2 + 2\alpha_i\lambda_t + 2\alpha_i u_{it} + 2\lambda_t u_{it}\right) \\
&= \mathrm{E}\left(\alpha_i^2 + \lambda_t^2 + u_{it}^2\right) = \sigma_\alpha^2 + \sigma_\lambda^2 + \sigma_u^2.
\end{aligned} \tag{4.16}$$

In the following, we assume $\lambda_t = 0$ and consider only unit specific error terms α_i

$$v_i' = (v_{i1} \dots v_{iT}), \quad v_{it} = \alpha_i + u_{it}. \tag{4.17}$$

To demonstrate the resulting dependencies in the covariance matrix of the error vector v_i', we look at the most simple $n = 2, T = 2$ case. With

$$\mathbf{v} = \begin{bmatrix} \alpha_1 + u_{11} \\ \alpha_1 + u_{12} \\ \alpha_2 + u_{21} \\ \alpha_2 + u_{22} \end{bmatrix}$$

and

$$\mathrm{Cov}(\mathbf{v}) = \mathrm{E}\left[\mathbf{v} - \mathrm{E}(\mathbf{v})\right]^2 = \mathrm{E}\left[\mathbf{v}\right]^2 = \mathrm{E}\left[\mathbf{v}\mathbf{v}'\right] \tag{4.18}$$

we get

$$\mathrm{Cov}(\mathbf{v}) = \mathrm{E}\left[\begin{pmatrix} \alpha_1 + u_{11} \\ \alpha_1 + u_{12} \\ \alpha_2 + u_{21} \\ \alpha_2 + u_{22} \end{pmatrix} \begin{pmatrix} \alpha_1 + u_{11} & \alpha_1 + u_{12} & \alpha_2 + u_{21} & \alpha_2 + u_{22} \end{pmatrix}\right]$$

$$= \mathrm{E}\begin{bmatrix} (\alpha_1 + u_{11})^2 & (\alpha_1 + u_{11})(\alpha_1 + u_{12}) & (\alpha_1 + u_{11})(\alpha_2 + u_{21}) & (\alpha_1 + u_{11})(\alpha_2 + u_{22}) \\ (\alpha_1 + u_{12})(\alpha_1 + u_{11}) & (\alpha_1 + u_{12})^2 & (\alpha_1 + u_{12})(\alpha_2 + u_{21}) & (\alpha_1 + u_{21})(\alpha_2 + u_{22}) \\ (\alpha_2 + u_{21})(\alpha_1 + u_{11}) & (\alpha_2 + u_{21})(\alpha_1 + u_{12}) & (\alpha_2 + u_{21})^2 & (\alpha_2 + u_{21})(\alpha_2 + u_{22}) \\ (\alpha_2 + u_{22})(\alpha_1 + u_{11}) & (\alpha_2 + u_{22})(\alpha_1 + u_{12}) & (\alpha_2 + u_{22})(\alpha_2 + u_{21}) & (\alpha_2 + u_{22})^2 \end{bmatrix} \tag{4.19}$$

Using the assumed property $\mathrm{E}(\alpha, u) = 0$, we find

$$\mathrm{Cov}(\mathbf{v}) = \mathrm{E}\begin{bmatrix} \alpha_1\alpha_1 + u_{11}u_{11} & \alpha_1\alpha_1 + u_{11}u_{12} & \alpha_1\alpha_2 + u_{11}u_{21} & \alpha_1\alpha_2 + u_{11}u_{22} \\ \alpha_1\alpha_1 + u_{12}u_{11} & \alpha_1\alpha_1 + u_{12}u_{12} & \alpha_1\alpha_2 + u_{12}u_{21} & \alpha_1\alpha_2 + u_{12}u_{22} \\ \alpha_2\alpha_1 + u_{21}u_{11} & \alpha_2\alpha_1 + u_{21}u_{12} & \alpha_2\alpha_2 + u_{21}u_{21} & \alpha_2\alpha_2 + u_{21}u_{22} \\ \alpha_2\alpha_1 + u_{22}u_{11} & \alpha_2\alpha_1 + u_{22}u_{12} & \alpha_2\alpha_2 + u_{22}u_{21} & \alpha_2\alpha_2 + u_{22}u_{22} \end{bmatrix} \tag{4.20}$$

Given the assumptions (see Eq. 4.15)

$$\mathrm{E}(u_{is}, u_{jt}) = \begin{cases} \sigma_u^2, \text{ if } i = j \text{ and } s = t \\ 0, \text{ else} \end{cases}$$

$$\mathrm{E}(\alpha_i, \alpha_j) = \begin{cases} \sigma_\alpha^2, \text{ if } i = j \\ 0, \text{ else} \end{cases}$$

we obtain

$$\mathrm{Cov}(\mathbf{v}) = \mathrm{E}\begin{bmatrix} \sigma_\alpha^2 + \sigma_u^2 & \sigma_\alpha^2 & 0 & 0 \\ \sigma_\alpha^2 & \sigma_\alpha^2 + \sigma_u^2 & 0 & 0 \\ 0 & 0 & \sigma_\alpha^2 + \sigma_u^2 & \sigma_\alpha^2 \\ 0 & 0 & \sigma_\alpha^2 & \sigma_\alpha^2 + \sigma_u^2 \end{bmatrix}. \tag{4.21}$$

Picking one specific unit i in the simple $n = 2,\ T = 2$ case, we find

$$\mathrm{Cov}(\mathbf{v}_i) = \begin{bmatrix} \sigma_\alpha^2 + \sigma_u^2 & \sigma_\alpha^2 \\ \sigma_\alpha^2 & \sigma_\alpha^2 + \sigma_u^2 \end{bmatrix} \tag{4.22}$$

with its inverse

$$\begin{aligned} [\mathrm{Cov}(\mathbf{v}_i)]^{-1} &= \begin{bmatrix} \sigma_\alpha^2 + \sigma_u^2 & \sigma_\alpha^2 \\ \sigma_\alpha^2 & \sigma_\alpha^2 + \sigma_u^2 \end{bmatrix}^{-1} \\ &= \begin{bmatrix} \frac{1}{\sigma_u^2} - \frac{\sigma_\alpha^2}{\sigma_u^2(\sigma_u^2+T\sigma_\alpha^2)} & -\frac{\sigma_\alpha^2}{\sigma_u^2(\sigma_u^2+T\sigma_\alpha^2)} \\ -\frac{\sigma_\alpha^2}{\sigma_u^2(\sigma_u^2+T\sigma_\alpha^2)} & \frac{1}{\sigma_u^2} - \frac{\sigma_\alpha^2}{\sigma_u^2(\sigma_u^2+T\sigma_\alpha^2)} \end{bmatrix} \end{aligned} \tag{4.23}$$

This inverted covariance matrix can also be written as

$$V_i^{-1} = \frac{1}{\sigma_u^2}\left[\left(I_T - \frac{1}{T}\mathbf{ee}'\right) + \psi\frac{1}{T}\mathbf{ee}'\right] = \frac{1}{\sigma_u^2}\left[Q + \psi\frac{1}{T}\mathbf{ee}'\right] \tag{4.24}$$

where

$$\psi = \frac{\sigma_u^2}{\sigma_u^2 + T\sigma_\alpha^2} \tag{4.25}$$

making use of the idempotent matrix

$$Q = I_T - \frac{1}{T}\mathbf{ee}' \tag{4.26}$$

with $\mathbf{e}$ being a vector with ones and dimension $T \times 1$. Q has the following properties

$$Q = Q' = Q^{-1} = QQ \tag{4.27}$$

and premultiplying a vector by Q results in a demeaned vector.

Given the following equation

$$\begin{aligned} V_i^{-1} = [\mathrm{Cov}(\mathbf{v}_i)]^{-1} &= \frac{1}{\sigma_u^2}\left(Q + \psi\frac{1}{T}\mathbf{ee}'\right) = \frac{1}{\sigma_u^2}\left(\left\{I_T - \frac{1}{T}\mathbf{ee}'\right\} + \psi\frac{1}{T}\mathbf{ee}'\right) \\ &= \frac{1}{\sigma_u^2}\left(I_T - \begin{bmatrix} \frac{1}{T} & \frac{1}{T} \\ \frac{1}{T} & \frac{1}{T} \end{bmatrix} + \begin{bmatrix} \frac{1}{T}\psi & \frac{1}{T}\psi \\ \frac{1}{T}\psi & \frac{1}{T}\psi \end{bmatrix}\right) \\ &= \frac{1}{\sigma_u^2}\left(I_T - \begin{bmatrix} \frac{\sigma_u^2+T\sigma_\alpha^2}{T(\sigma_u^2+T\sigma_\alpha^2)} & \frac{\sigma_u^2+T\sigma_\alpha^2}{T(\sigma_u^2+T\sigma_\alpha^2)} \\ \frac{\sigma_u^2+T\sigma_\alpha^2}{T(\sigma_u^2+T\sigma_\alpha^2)} & \frac{\sigma_u^2+T\sigma_\alpha^2}{T(\sigma_u^2+T\sigma_\alpha^2)} \end{bmatrix} + \begin{bmatrix} \frac{1}{T}\frac{\sigma_u^2}{\sigma_u^2+T\sigma_\alpha^2} & \frac{1}{T}\frac{\sigma_u^2}{\sigma_u^2+T\sigma_\alpha^2} \\ \frac{1}{T}\frac{\sigma_u^2}{\sigma_u^2+T\sigma_\alpha^2} & \frac{1}{T}\frac{\sigma_u^2}{\sigma_u^2+T\sigma_\alpha^2} \end{bmatrix}\right) \\ &= \frac{1}{\sigma_u^2}\left\{I_T - \begin{bmatrix} \frac{\sigma_\alpha^2}{\sigma_u^2+T\sigma_\alpha^2} & \frac{\sigma_\alpha^2}{\sigma_u^2+T\sigma_\alpha^2} \\ \frac{\sigma_\alpha^2}{\sigma_u^2+T\sigma_\alpha^2} & \frac{\sigma_\alpha^2}{\sigma_u^2+T\sigma_\alpha^2} \end{bmatrix}\right\} = \frac{1}{\sigma_u^2}\left\{I_T - \frac{\sigma_\alpha^2}{(\sigma_u^2 + T\sigma_\alpha^2)}\mathbf{ee}'\right\} \end{aligned} \tag{4.28}$$

we have

$$\mathrm{Cov}(\mathbf{v}_i) = \mathrm{E}\left(\mathbf{v}_i\mathbf{v}_i'\right) = \sigma_u^2 I_T + \sigma_\alpha^2 \mathbf{e}\mathbf{e}' = V_i \tag{4.29}$$

and for the inverse

$$V_i^{-1} = \frac{1}{\sigma_u^2}\left[I_T - \frac{\sigma_\alpha^2}{\sigma_u^2 + T\sigma_\alpha^2}\mathbf{e}\mathbf{e}'\right] = \frac{1}{\sigma_u^2}\left[Q - \psi\frac{1}{T}\mathbf{e}\mathbf{e}'\right]. \tag{4.30}$$

Because $\mathrm{E}(v_{it}v_{is}) = \sigma_\alpha^2$, the error terms v_{it} are correlated within units. Therefore OLS estimation is inefficient and efficiency can be improved by taking the known structure of the covariance matrix into account.

4.2.3.2 The GLS Estimator for the Random Effects Model

In general notation, the OLS estimator is

$$\hat{\beta}^{\mathrm{OLS}} = \left(X'X\right)^{-1} X'y \tag{4.31}$$

and the generalized least squares estimator is

$$\hat{\beta}^{\mathrm{GLS}} = \left(X'V^{-1}X\right)^{-1} X'V^{-1}y. \tag{4.32}$$

In the case of unit specific random effects, the $(n \times T) \times (n \times T)$ inverse covariance matrix V is block-diagonal with identical blocks V_i at its diagonal (see above).

It is interesting that the random effects estimator can be obtained as a weighted linear combination of the within β^{W} and between estimators β^{B}:

$$\hat{\beta}^{\mathrm{GLS}} = \Delta\hat{\beta}^{\mathrm{B}} + (I_K - \Delta)\,\hat{\beta}^{\mathrm{W}}. \tag{4.33}$$

Here, the parameter vector β^{GLS} does not include the overall intercept, which can be obtained as

$$\hat{\beta}_0^{\mathrm{GLS}} = \bar{y} - \hat{\beta}^{\mathrm{GLS}\prime}\bar{x}. \tag{4.34}$$

The weighting matrix Δ is

$$\begin{aligned}\Delta = \psi T\left[\sum_{i=1}^{n} X_i'QX_i + \psi T\sum_{i=1}^{n}\left(\bar{x}_i - \bar{x}\right)\left(\bar{x}_i - \bar{x}\right)'\right]^{-1} \\ \times\left[\sum_{i=1}^{n} X_i'Qy_i + \psi T\sum_{i=1}^{n}\left(\bar{x}_i - \bar{x}\right)\left(\bar{y}_i - \bar{y}\right)\right].\end{aligned} \tag{4.35}$$

Note that ψ and therefore V_i is not known and has to be estimated to make the GLS-estimator feasible. The variance components $\hat{\sigma}_u^2$ and $\hat{\sigma}_\alpha^2$ can be estimated using the within and between estimators:

$$\hat{\sigma}_u^2 = \frac{\sum_{i=1}^{n} \left[(y_{it} - \bar{y}_i) - \hat{\beta}^{\mathrm{W}} (x_i - \bar{x}_i) \right]^2}{n(T-1) - K} \tag{4.36a}$$

$$\hat{\sigma}_\alpha^2 = \frac{\sum_{i=1}^{n} \left[\bar{y}_i - \tilde{\mu} - \beta^{\mathrm{B}\prime} \bar{x}_i \right]^2}{n - (K+1)} - \frac{1}{T} \hat{\sigma}_u^2. \tag{4.36b}$$

Based on the estimators of the variance components, an estimate $\hat{\psi}$ of the unknown parameter ψ can be obtained:

$$\hat{\psi} = \frac{\hat{\sigma}_u^2}{\hat{\sigma}_u^2 + T\hat{\sigma}_\alpha^2}. \tag{4.37}$$

An easy way to find the random effects estimator empirically is applying OLS to transformed variables $\breve{y}_{it}$ and $\breve{x}_{it}$ making use of $\hat{\psi}$ with

$$\breve{y}_{it} = y_{it} - \left(1 - \psi^{1/2}\right) \bar{y}_i \tag{4.38a}$$

$$\breve{x}_{it} = x_{it} - \left(1 - \psi^{1/2}\right) \bar{x}_i. \tag{4.38b}$$

4.2.3.3 The Maximum Likelihood Estimator

The GLS estimation is a two-step procedure. In the first step, within and between estimators are obtained ignoring the covariance structure. Based on these estimates the covariance matrix is estimated and used in the second step to estimate the feasible GLS estimator. One can argue that this procedure is not efficient, as the estimates in the first round ignore the dependence structure in the residuals.

One possibility is to iterate the procedure. The basic idea is that in the second round the estimation of residuals which are used to estimate the variance would be improved, resulting in an improved estimation of the parameter vector β and so on. A method that basically does this iteration is the maximum likelihood estimation. It is common to assume that the error terms are distributed normally.

The basic idea of the maximum likelihood estimation is to find the parameters which would give the observed values the highest probability to realize given the model assumptions.

In the univariate case the density function is

$$\varphi(X) = \frac{1}{\sigma\sqrt{2\pi}} \cdot \mathrm{e}^{-\frac{1}{2} \cdot \frac{(X-\mu)^2}{\sigma^2}} \tag{4.39}$$

Assuming independence between the different realizations, the ex ante probability of the realized values is

$$\begin{aligned} W(x_1 \cap x_2 \cap \ldots \cap x_n) &= \varphi(x_1) \cdot \varphi(x_2) \cdot \ldots \cdot \varphi(x_n) \\ &= L(\mu, \sigma | x_1, \ldots, x_n) \\ &= \left(\frac{1}{\sigma_X \sqrt{2\pi}} \right)^n \cdot \mathrm{e}^{-\frac{1}{2} \cdot \frac{\sum_{i=1}^{n} (x_i - \mu)^2}{\sigma^2}} . \end{aligned} \tag{4.40}$$

Since taking the logarithm is a strict monotone transformation, the likelihood and log-likelihood have their maxima at identical points. Usually, the log-likelihood function is easier to work with

$$\ln L(\mu, \sigma | x_1, \ldots, x_n) = -n \cdot \left(\ln \sigma + \ln \sqrt{2\pi} \right) - \frac{1}{2\sigma_2} \cdot \sum_{i=1}^{n} (x_i - \mu)^2. \tag{4.41}$$

In the multivariate case μ is a vector and instead of the variance parameter (scalar) we have to consider the covariance matrix Σ. The product of densities in the multivariate case is

$$W(x_1 \cap x_2 \cap \ldots \cap x_n) = (2\pi)^{-n/2} |\boldsymbol{\Sigma}|^{-1/2} \mathrm{e}^{(-1/2)(\mathbf{x}-\boldsymbol{\mu})' \boldsymbol{\Sigma}^{-1} (\mathbf{x}-\boldsymbol{\mu})} . \tag{4.42}$$

We now return to our random effects regression model using β_0 for the overall intercept and β for the vector of slope coefficients. Despite the index i at the variance matrix in the summation across firms, in this model the variance matrix is identical for all firms i. In the multivariate case the likelihood for the vector of residuals is

$$L(\beta, \Sigma | x) = (2\pi)^{-nT/2} |V|^{-n/2} \, \mathrm{e}^{-\frac{1}{2} \sum_{i=1}^{n} (\mathbf{y}_i - \mathbf{e}\beta_0 - X_i\beta)' V_i^{-1} (\mathbf{y}_i - \mathbf{e}\beta_0 \mu - X_i\beta)} \tag{4.43}$$

and the log-likelihood is

$$\begin{aligned} \ln L = &- \frac{nT}{2} \ln 2\pi - \frac{n}{2} \ln |V_i| \\ &- \frac{1}{2} \sum_{i=1}^{n} (\mathbf{y}_i - \mathbf{e}\beta_0 - X_i\beta)' \, V_i^{-1} \, (y_i - \mathbf{e}\beta_0 - X_i\beta) \end{aligned} \tag{4.44}$$

with covariance matrix

$$V_i = \sigma_u^2 I_T + \sigma_\alpha^2 \mathbf{e}\mathbf{e}'. \tag{4.45}$$

Inserting the inverse of the covariance matrix for units

$$V_i^{-1} = \frac{1}{\sigma_u^2}\left[\left(I_T - \frac{1}{T}\mathbf{ee}'\right) + \psi\frac{1}{T}\mathbf{ee}'\right] = \frac{1}{\sigma_u^2}\left[Q + \frac{\sigma_u^2}{\sigma_u^2 + T\sigma_\alpha^2}\frac{1}{T}\mathbf{ee}'\right] \tag{4.46}$$

and using the determinant of the covariance matrix

$$|V_i| = \sigma_u^{2(T-1)}\left(\sigma_u^2 + T\sigma_\alpha^2\right) \tag{4.47}$$

results in the following log-likelihood

$$\begin{aligned}
\ln L = &-\frac{nT}{2}\ln 2\pi - \frac{n}{2}\ln \sigma_u^{2(T-1)}\left(\sigma_u^2 + T\sigma_\alpha^2\right) \\
&-\frac{1}{2}\sum_{i=1}^{n}(y_i - \mathbf{e}\beta_0 - X_i\beta)' \\
&\cdot\left(\frac{1}{\sigma_u^2}\left[Q + \frac{\sigma_u^2}{\sigma_u^2 + T\sigma_\alpha^2}\frac{1}{T}\mathbf{ee}'\right]\right)(y_i - \mathbf{e}\beta_0 - X_i\beta) \\
= &-\frac{nT}{2}\ln 2\pi - \frac{n}{2}\ln\left(\sigma_u^{2(T-1)}\right) - \frac{n}{2}\ln\left(\sigma_u^2 + T\sigma_\alpha^2\right) \\
&-\frac{1}{2\sigma_u^2}\sum_{i=1}^{n}(y_i - \mathbf{e}\beta_0 - X_i\beta)'Q\,(y_i - \mathbf{e}\beta_0 - X_i\beta) \\
&-\frac{1}{T}\frac{1}{\sigma_u^2 + T\sigma_\alpha^2}\sum_{i=1}^{n}(y_i - \mathbf{e}\beta_0 - X_i\beta)'\,ee'\,(y_i - \mathbf{e}\beta_0 - X_i\beta)\,.
\end{aligned} \tag{4.48}$$

Because of

$$\frac{1}{T}(y_i - \mathbf{e}\beta_0 - X_i\beta)'\,\mathbf{e} = \bar{y}_i - \beta_0 - \beta'\bar{x}_i \tag{4.49}$$

and

$$\frac{1}{T}\mathbf{e}'\,(y_i - \mathbf{e}\beta_0 - X_i\beta) = \bar{y}_i - \mu - \beta'\bar{x}_i \tag{4.50}$$

we finally obtain

$$\begin{aligned}
\ln L = &-\frac{nT}{2}\ln 2\pi - n\,(T-1)\ln\left(\sigma_u^2\right) - \frac{n}{2}\ln\left(\sigma_u^2 + T\sigma_\alpha^2\right) \\
&-\frac{1}{2\sigma_u^2}\sum_{i=1}^{n}(y_i - \mathbf{e}\beta_0 - X_i\beta)'\,Q\,(y_i - \mathbf{e}\beta_0 - X_i\beta) \\
&-\frac{T}{\sigma_u^2 + T\sigma_\alpha^2}\sum_{i=1}^{n}\left(\bar{y}_i - \beta_0 - \beta'\bar{x}_i\right)^2.
\end{aligned} \tag{4.51}$$

The value of the log-likelihood will be maximized by means of the parameter vector

$$\delta' = \left(\beta_0, \beta', \sigma_u^2, \sigma_\alpha^2\right). \tag{4.52}$$

For the numerical optimization, we can make use of the analytic derivatives of the log-likelihood

$$\frac{\partial \ln L}{\partial \mu} = \frac{T}{\sigma_u^2 + T\sigma_\alpha^2} \sum_{i=1}^{n} \left(\bar{y}_i - \beta_0 - \beta'\bar{x}_i\right)^2 = 0 \tag{4.53a}$$

$$\frac{\partial \ln L}{\partial \beta} = \frac{1}{\sigma_u^2} \left(\begin{array}{c} \sum_{i=1}^{n} \left(y_i - \mathbf{e}\beta_0 - X_i\beta\right)' QX_i \\ -\frac{T\sigma_u^2}{\sigma_u^2 + T\sigma_\alpha^2} \sum_{i=1}^{n} \left(\bar{y}_i - \beta_0 - \beta'\bar{x}_i\right)^2 \end{array} \right) = 0 \tag{4.53b}$$

$$\begin{aligned} \frac{\partial \ln L}{\partial \sigma_u^2} = &- \frac{n\,(T-1)}{2\sigma_u^2} - \frac{n}{2\left(\sigma_u^2 + T\sigma_\alpha^2\right)} \\ &+ \frac{1}{2\sigma_u^4} \sum_{i=1}^{n} \left(y_i - \mathbf{e}\beta_0 - X_i\beta\right)' Q \left(y_i - \mathbf{e}\beta_0 - X_i\beta\right) \\ &+ \frac{T}{2\left(\sigma_u^2 + T\sigma_\alpha^2\right)^2} \sum_{i=1}^{n} \left(\bar{y}_i - \beta_0 - \beta'\bar{x}_i\right)^2 = 0 \end{aligned} \tag{4.53c}$$

$$\frac{\partial \ln L}{\partial \sigma_\alpha^2} = - \frac{nT}{2\left(\sigma_u^2 + T\sigma_\alpha^2\right)} + \frac{T^2}{2\left(\sigma_u^2 + T\sigma_\alpha^2\right)^2} \sum_{i=1}^{n} \left(\bar{y}_i - \beta_0 - \beta'\bar{x}_i\right)^2 = 0. \tag{4.53d}$$

4.2.4 Fixed or Random?

The choice between fixed and random effects models can either be grounded in substantial or technical considerations. In general, it is rather difficult to provide any substantial arguments for the random processes typically assumed in econometrics. Taken literally, the observed values of statistical variables characterizing objects have to be regarded as realizations obtained by using a random number generator. Of course, this is obviously not the case when analyzing empirical data and the usual argument is that observations look somewhat similar to realizations generated using a random generator. In our context of production functions, one may argue that a large number of influences not captured in the regression function will sum up to something that can be fruitfully seen as a stochastic error term. Turning to the question of fixed or random, we can speculate about the nature of persistent deviations α_i from the "normal" production relation. We can also ask whether we are interested in the individual production processes of the individual firms (fixed α_i)

or whether we are rather interested in an "average behavior" seeing the individual units only as providing bits of information to this "average behavior" (random α_i). These aspects are formally expressed in the way we formulate expected values of output. Do we want to estimate the expected output given inputs for a specific unit i, that is $E[Y_i|X_i;\alpha_i]$, or are we rather interested in the expected output given inputs for any firm (or average firms) $E[Y_i|X_i]$?

While discussions about the presumed stochastic nature of statistical variables are rather sparse, more often purely formal arguments are provided. Because the random effects estimator is biased if individual effects α_i are correlated with inputs x_i', that is $\text{Cov}[x_i', \alpha_i] \neq 0$ (e.g., if bigger firms have smaller values of α_i), the decision is simply based on the formal question whether the empirical data hint for a "significant" correlation (note that in empirical data there is always correlation). If there is "significant evidence" for correlation, the fixed effects estimator is chosen which is unbiased in this case. If there is no "significant evidence" for correlation, the random effects estimator is chosen, because in the absence of correlation between α_i and x_i', both estimators are unbiased, but the random effects estimator is more efficient. Usually, the Hausman-test is applied to decide about whether the correlation contained in the data is "significant".

4.3 Panel Data Analysis with R

4.3.1 Generating a Panel Data Set

We start with creating a small panel data set with three units ($i = 1, 2, 3$) observed at three different time periods ($t = 1, 2, 3$). The way the data are generated meets the assumption of a model with normally distributed idiosyncratic errors u_{it}, normally distributed random effects for units α_i, and normally distributed random effects for periods λ_t:

```
N <- 3
M <- 3
NM <- N*M
set.seed(123)
x <- runif(NM)
a <- rep(rnorm(N,sd=0.2),each=M)
l <- rep(rnorm(N,sd=0.2),each=M)
u <- rnorm(NM,sd=0.2)
y <- 1+a+l+x+u
i <- rep(1:3,each=3)
m <- rep(1:3,3)
```

The two vectors i and m contain the unit and time index respectively and will turn out helpful when carrying out panel data specific data transformations. To make evident which data belong to a single unit i, we use triangles, dots, and rhombuses for units 1,2,3. To indicate the time periods 1,2,3, we use three different sizes, growing with t. In Fig. 4.1, we show the estimated pooled regression, the three

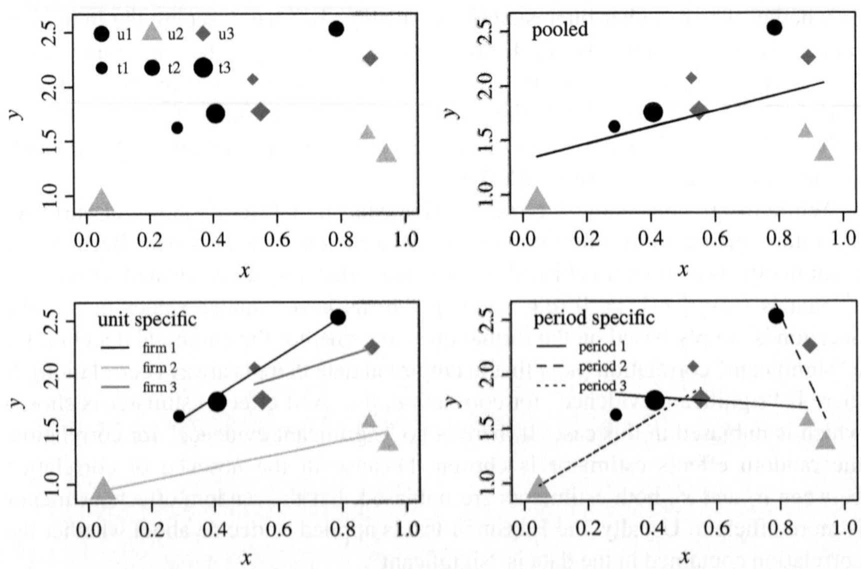

Fig. 4.1 Panel data, pooled, unit, and time specific regressions

separately estimated time series regressions for the three units and three separately estimated cross-sectional regressions for time periods 1, 2, and 3. These regressions can be obtained with the function `lm()`.

```
ylimits <- range(y)
xlimits <- range(x)
picture <- function(a) {
  plot(x[1:3],y[1:3],xlim=c(0,1),ylim=c(0.9,2.6),
       xlab="x",ylab="y",pch=16,cex=c(1.5,2,2.5))
  points(x[4:6],y[4:6],pch=17,
         cex=c(1.5,2,2.5),col='grey')
  points(x[7:9],y[7:9],pch=18,
         cex=c(1.5,2,2.5),col='grey40')
}
# data points
picture(1)
legend(0, 2.7, c("u1","t1","u2","t2","u3","t3"),
       pch=c(16,16,17,16,18,16),
       col=c(1,1,"grey",1,"grey40",1),
       pt.cex=c(2,1.5,2,2,2,2.5), cex=0.8,
       ncol=3, bty="n", y.intersp=1.5)
# pooled
picture(1)
reg.pooled <- lm(y~x) #;reg.pooled
lines(x,fitted(reg.pooled),lwd=1)
text(0,2.5,"pooled",pos=4)
# unit specific
picture(1)
```

```
nr <- 1:9
colv <- c(1,"grey","grey40")
lty <- 1:3
for (j in 1:3){
  ind <- nr[i==j]
  reg <- lm(y[ind]~x[ind])
  lines(x[ind],fitted(reg),lwd=1.5,col=colv[j])
  }
text(0,2.5,"unit specific",pos=4)
legend(0,2.44,c("firm 1","firm 2","firm 3"),
       lwd=1.5,lty=1,col=colv,cex=0.7,bty="n",seg.len=2)
# period specific
picture(1)
for (j in 1:3){
  ind <- nr[m==j]
  reg <- lm(y[ind]~x[ind])
  lines(x[ind],fitted(reg),lty=j,lwd=1.5)
  }
text(0,2.5,"period specific",pos=4)
legend(0,2.44,c("period 1","period 2","period 3"),
       lwd=1.5,lty=1:3, cex=0.7, bty="n", seg.len=2)
```

We now turn to the estimation of fixed effects models. Again, these regressions can be obtained with the function `lm()`. Note that there is a difference between estimating separate regressions and one model with fixed effects, as the variance of the residual in one model with fixed effects is restricted to be identical for all units/time periods. In the following, we estimate in turn the pooled model, the model with fixed unit effects (slopes identical but intercepts varying across units), the model with fixed time effects (slopes identical but intercepts varying across time periods), and a model with fixed unit and time effects (but identical slope parameters).

```
fi <- as.factor(i)
fm <- as.factor(m)
# pooled
reg.pooled <- lm(y~x); reg.pooled

##
## Call:
## lm(formula = y ~ x)
##
## Coefficients:
## (Intercept)            x
##       1.316        0.754

# fixed unit effects
reg.fixed.units <- lm(y~x+fi); reg.fixed.units

##
## Call:
## lm(formula = y ~ x + fi)
##
```

```
## Coefficients:
## (Intercept)           x         fi2         fi3
##      1.5369      0.8710     -0.7932     -0.0784

# fixed time effects
reg.fixed.time <- lm(y~x+fm); reg.fixed.time

##
## Call:
## lm(formula = y ~ x + fm)
##
## Coefficients:
## (Intercept)           x         fm2         fm3
##       1.541       0.370       0.184      -0.179

# fixed unit and time effects
reg.fixed.both <- lm(y~x+fi+fm); reg.fixed.both

##
## Call:
## lm(formula = y ~ x + fi + fm)
##
## Coefficients:
## (Intercept)           x         fi2         fi3
##      1.6713      0.6086     -0.7596     -0.0358
##         fm2         fm3
##      0.1108     -0.1242
```

4.3.2 Some Transformations with Panel Data

When working with data sets containing a large number of units, the simple approach using many (sometimes thousands of) dummy variables may lead to computational difficulties. In this case, demeaning the data before running the regression might be appropriate. This could be accomplished using the `tapply()`-command.

We illustrate this by estimating the slope parameter of the model with fixed unit effects. Note that the regression coefficient obtained is identical to that obtained with dummy variables. But as the computer cannot know that we already "estimated" three unit specific means for demeaning the data, the standard errors would have to be corrected to allow sensible inference.

```
xb <- unlist(tapply(x,i,function(z) z-mean(z)))
yb <- unlist(tapply(y,i,function(z) z-mean(z)))
lm(yb~xb)$coef

## (Intercept)          xb
##   1.612e-17   8.710e-01
```

In a similar manner, we could subtract time period specific means to estimate the model with fixed time effects with demeaned data. Accordingly, the slope parameter of the model with fixed unit and fixed time effects can be obtained after a twofold demeaning. In this case, the overall mean has to be added after the twofold demeaning.

```
xbb <- unlist(tapply(xb,
                     m,function(z) z-mean(z)))+mean(x)
ybb <- unlist(tapply(yb,
                     m,function(z) z-mean(z)))+mean(y)
lm(ybb~xbb)$coef

## (Intercept)         xbb
##      1.4016      0.6086
```

Note that the demeaned data (for unit specific levels) contain only information about deviations from unit specific levels whereas information on the differences in average levels is discarded. If there had been no variation across time, all differences in levels between units would be removed by the demeaning and the resulting data would contain no information at all.

The opposite happens with the between transformation. Using averages across time periods means that only the information about differences in levels between units is regarded. The number of observations is reduced from $n \cdot T$ towards n and all variation within units across time is discarded.

```
xm <- tapply(x,i,mean)
ym <- tapply(y,i,mean)
lm(ym~xm)$coef

## (Intercept)          xm
##       2.456      -1.173
```

4.3.3 Estimating Panel Data Models Using the plm Package

Handling panel data is comfortable with the panel data package plm. The R-package "plm" (`library(plm)`) contains a variety of panel data methods and several panel data sets. Beside our small artificial panel data set, we will use the data set "EmpUK" available in the package which has been studied by Arellano and Bond (1991).

Figure 4.2 shows the three time series regressions for the models defined in Sect. 4.3.1.

```
# left panel
picture(1)
for (j in 1:3){
  ind <- nr[i==j]
  lines(x[ind],fitted(reg.fixed.units)[ind],
        lwd=1,col=colv[j])
}
```

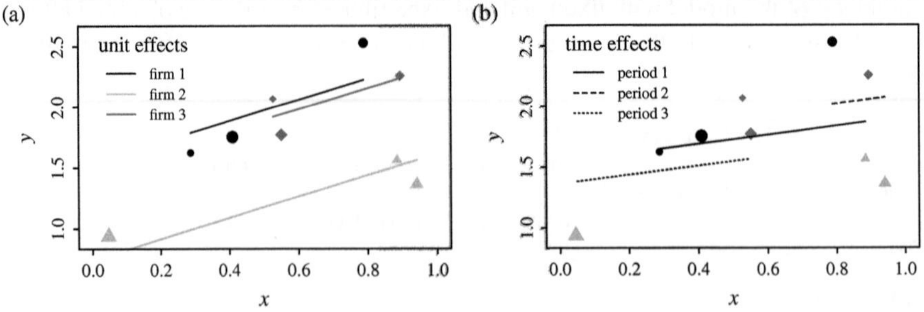

Fig. 4.2 Fixed effects. (**a**) Fixed unit effects. (**b**) Fixed time effects

```
text(0,2.5,"unit effects",pos=4,cex=1.4)
legend(0,2.45,c("firm 1","firm 2","firm 3"),
       lty=1,col=colv,bty="n",seg.len=2)
# right panel
picture(1)
lty <- c(1,2,3)
for (j in 1:3){
  ind <- nr[m==j]
  lines(x[ind],fitted(reg.fixed.time)[ind],lty=lty[j])
  }
text(0,2.5,"time effects",pos=4,cex=1.4)
legend(0,2.45,c("period 1","period 2","period 3"),
       lty=1:3,bty="n",seg.len=2)
```

The package plm uses a specific kind of dataframe containing additional information of the panel structure of the data in the form of an index attribute. The command `pdata.frame` transforms a dataframe into a panel dataframe (pdataframe for short).

```
library(plm)
d <- pdata.frame(data.frame(i,m,x,y));d[1:2,]

##     i m      x     y
## 1-1 1 1 0.2876 1.623
## 1-2 1 2 0.7883 2.528
```

The plm-package contains some commands useful to transform panel data. The simple alternative for our already discussed demeaning is the command `Within()`. The command `between()` returns unit specific means across time. The command corresponding to linear models `lm()` for panel data models is `plm()`. The model option allows (inter alia) to choose between pooled, fixed, and random models. The option effect allows to specify whether the fixed effects are for units ("individual"), time periods ("time"), or both ("twoways").

The R commands for our small artificial data set are as follows:

```
library(plm)
# fixed unit effects
plm(y~x, data=d, model="within",
    effect="individual")$coef

##     x
## 0.871

# fixed unit and time effects
plm(y~x, data=d, model="within",
    effect="twoways")$coef

##      x
## 0.6086

# between model for units
plm(y~x, data=d, model="between",
    effect="individual")$coef

## (Intercept)           x
##       2.456      -1.173

# random unit effects
plm(y~x, data=d, model="random",
    effect="individual")$coef

## (Intercept)           x
##      1.2503      0.8644

# random unit and time effects
plm(y~x, data=d, model="random",
    effect="twoways")$coef

## (Intercept)           x
##      1.2521      0.8614
```

Note that despite the fact that the last model (two-ways random effects) corresponds to the generation of our small artificial dataframe, we obtain a negative estimate of the variance of the random time effect. In plm there are five different random estimators implemented. The default estimator has been proposed by Swamy. Finally, we try to estimate a Cobb–Douglas production function using the Arellano–Bond data set contained in the plm package. We load the data set and construct a pdataframe. It covers the 140 firms observed in the time period 1976 to 1984. The panel data set is unbalanced as firms have been observed between 7 and 9 years. We use employment and capital stock as inputs and the net value added at fixed factor costs as output. We estimate a model with fixed unit and time effects.

```
library(plm)
data("EmplUK", package="plm")
E <- pdata.frame(EmplUK)
reg <- plm(log(output) ~ log(emp)+log(capital), data=E,
           model="within",effect="twoways")
```

```
summary(reg)

## Twoways effects Within Model
##
## Call:
## plm(formula = log(output) ~ log(emp) + log(capital), data = E,
##     effect = "twoways", model = "within")
##
## Unbalanced Panel: n=140, T=7-9, N=1031
##
## Residuals :
##     Min.   1st Qu.   Median  3rd Qu.     Max.
## -0.13400 -0.03680  0.00132  0.03680  0.18500
##
## Coefficients :
##              Estimate Std. Error t-value Pr(>|t|)
## log(emp)       0.0341     0.0137    2.50    0.013 *
## log(capital)   0.0470     0.0116    4.05 0.000056 ***
## ---
## Signif. codes:
## 0 '***' 0.001 '**' 0.01 '*' 0.05 '.' 0.1 ' ' 1
##
## Total Sum of Squares:    2.63
## Residual Sum of Squares: 2.45
## R-Squared       :  0.0677
##       Adj. R-Squared :  0.0578
## F-statistic: 31.972 on 2 and 881 DF, p-value: 3.94e-14
```

Note that the model information given by the command summary() refers to the twofold demeaned data. This results in a very low fraction of explained variance (about 6 %). Most variance in the dependent variable is eliminated by the twofold demeaning before estimation. If we estimate the model with dummy variables for firms and years, the fraction of explained variance is about 73 %.

One should also keep in mind that these models only provide some descriptive information about the data. As the empirical data have not been generated using a simple random generator, estimated models at best can provide some guidance for further speculations about empirical production processes. While panel data models have the advantage of providing highly condensed descriptive information, this is practically always gained at the cost of false assumptions. For example the assumption of identical slope parameters across firms is presumably false as the figure of simple one-factor production functions reveals. In Fig. 4.3, we have drawn the 140 firm specific linear one-factor production functions into the scatter plot. In Fig. 4.4, we show the distribution of the firm specific slope coefficients of the two-factor Cobb–Douglas production function.

```
ly <- log(E$output)
lemp <- log(E$emp)
lcap <- log(E$capital)
lf <- length(unique(E$firm))
erge <- rep(NA,lf)
```

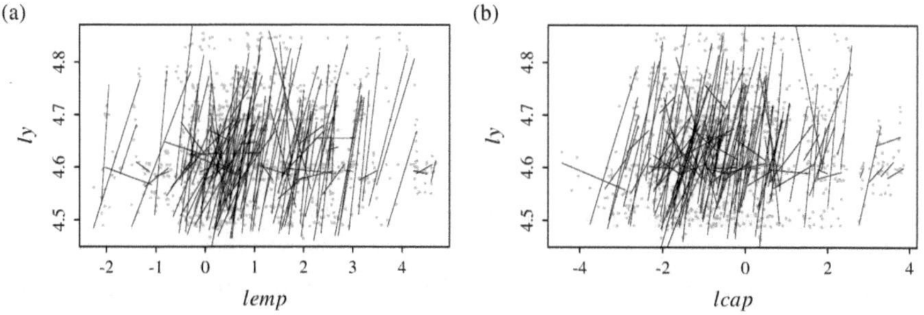

Fig. 4.3 Firm specific linear one-factor production functions. (**a**) For labor. (**b**) For capital

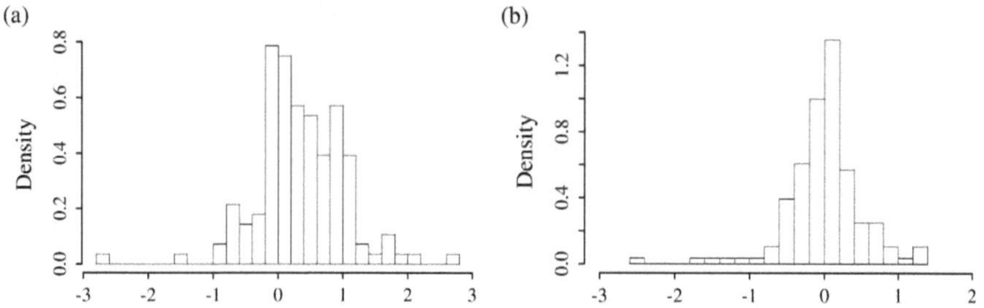

Fig. 4.4 Distribution of firm specific slope coefficients. (**a**) Log-employment. (**b**) Log-capital

```
ergc <- erge
plot(lemp,ly,pch=19,cex=0.5,col="darkgrey")
for (i in 1:lf){
  ind <- E$firm==i
  lmi <- lm(ly[ind]~lemp[ind])
  lines(lemp[ind],fitted(lmi))
  erge[i] <- lmi$coef[2]
  }
plot(lcap,ly,pch=19,cex=0.5,col="darkgrey")
for (i in 1:lf){
  ind <- E$firm==i
  lmi <- lm(ly[ind]~lcap[ind])
  lines(lcap[ind],fitted(lmi))
  ergc[i] <- lmi$coef[2]
  }
```

```
for (i in 1:lf){
  ind <- E$firm==i
  lmi <- lm(ly[ind]~lemp[ind]+lcap[ind])
  erge[i] <- lmi$coef[2]
  ergc[i] <- lmi$coef[3]
  }
```

```
hist(erge,nclass=20,main="",prob=T,ylab="Density",
     xlab="log-employment")
hist(ergc,nclass=20,main="",prob=T,ylab="Density",
     xlab="log-capital", xlim=c(-3,2))
```

4.3.4 Dynamic Panel Data Models

In the last decades it has become a standard practice to formulate and estimate dynamic panel data models, i.e., the model includes the production of the previous period (year) as an explanatory variable. Obviously, it is rather difficult to give such a model any economic meaning, as it states that without using any labor and capital in period t, nevertheless there will be output in period t just because there has been some output in period $t-1$. One argument for a dynamic specification of the production function might be that the production period for some goods may be of considerable length and therefore may extend towards the next period. In this case a fraction of the production which started in year $t-1$ may continue in period t. This will result in autocorrelated output data.

Empirically, we usually find a considerable correlation between the production in $t-1$ and t. To exemplify this, we use again the Arellano and Bond data for 140 firms operating in the UK. We draw the 140 linear regressions between logarithmic output in period $t-1$ (y_{t-1}) and in period t (y_t) and show the empirical distribution of the 140 individual autoregression coefficients (see Fig. 4.5).

```
library(plm)
data("EmplUK", package="plm")
E <- pdata.frame(EmplUK)
names(E)[6:7] <- c("cap","y")
ly <- log(E$y)
ly_1 <- lag(ly,1)
lf <- length(unique(E$firm))
erge <- rep(NA,lf)
plot(ly_1,ly,pch=19,cex=0.5,col="darkgrey")
```

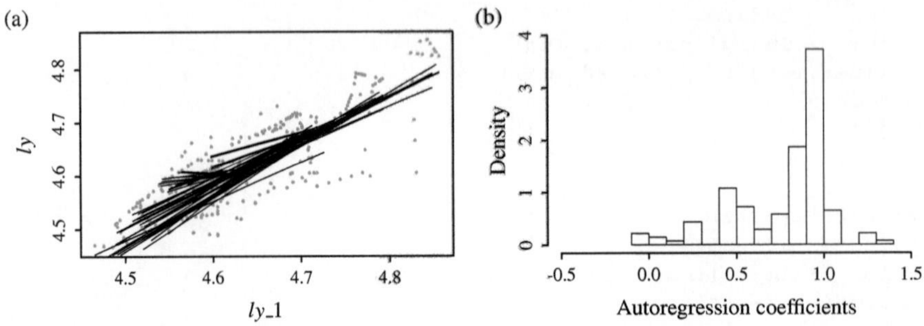

Fig. 4.5 Firm specific autoregression. (**a**) Linear regressions. (**b**) Individual autoregression coefficients

```
for (i in 1:lf){
  ind <- E$firm==i
  lmi <- lm(ly[ind]~ly_1[ind])
  lines(na.omit(ly_1[ind]),fitted(lmi))
  erge[i] <- lmi$coef[2]
  }
hist(erge,nclass=20,main="",prob=T,ylab="Density",
     xlab="Autoregression coefficients",
     xlim=c(-0.5,1.5))
```

We restrict the discussion to one specific dynamic model which includes a firm specific effect

$$y_{it} = \rho y_{i,t-1} + x_{it}'\beta_1 + \alpha_i + v_{it}. \tag{4.54}$$

Note that firmwise demeaning to eliminate the fixed effect α_i results in a correlation between the error term $v_{it} - \bar{v}_i$ and $y_{i,t-1} - \bar{y}_{i,-1}$. For example v_{it-1} is contained in $y_{i,t-1}$ and with weight $1/T$ also in $\bar{v}_i$.

An alternative to demeaning is taking differences

$$\begin{aligned} y_{it} - y_{i,t-1} &= \rho\,(y_{i,t-1} - y_{i,t-2}) + (x_{it}' - x_{i,t-1}')\beta_1 + (v_{it} - v_{i,t-1}) \\ \Delta y_{it} &= \rho\Delta y_{i,t-1} + \Delta x_{it}'\beta_1 + \Delta v_{i,t-1}. \end{aligned} \tag{4.55}$$

Again the estimation with OLS would be inconsistent as the regressor $\Delta y_{i,t-1}$ is correlated with Δv_{it} because $v_{i,t-1}$ is contained in both terms. Ignoring the negative correlation of $\Delta y_{i,t-1}$ and Δv_{it} we obtain the following parameter estimates:

```
reg3 <- plm(diff(log(y)) ~ lag(diff(log(y)),1)
            +diff(log(emp))+diff(log(cap)),
            model="pooling",effect=NULL,data=E)
round(summary(reg3)$coef,4)

##                        Estimate Std. Error t-value
## (Intercept)             -0.0185     0.0021 -8.6399
## lag(diff(log(y)), 1)     0.2180     0.0374  5.8289
## diff(log(emp))           0.0553     0.0163  3.4023
## diff(log(cap))           0.0103     0.0135  0.7624
##                        Pr(>|t|)
## (Intercept)              0.0000
## lag(diff(log(y)), 1)     0.0000
## diff(log(emp))           0.0007
## diff(log(cap))           0.4461
```

One possible way to circumvent this problem is the use of instruments. The idea is to find a variable (or several variables) which is strongly correlated with the regressor $\Delta y_{i,t-1}$ but uncorrelated with the error term $\Delta v_{i,t}$. The fitted values of the regression of the regressor on the instrument will then be uncorrelated with the error term and hence OLS will be consistent. As a simple example we use $\Delta y_{i,t-2}$ as instrument for $\Delta y_{i,t-1}$ (cf. Anderson and Hsiao (1981)) in a simple Cobb–Douglas

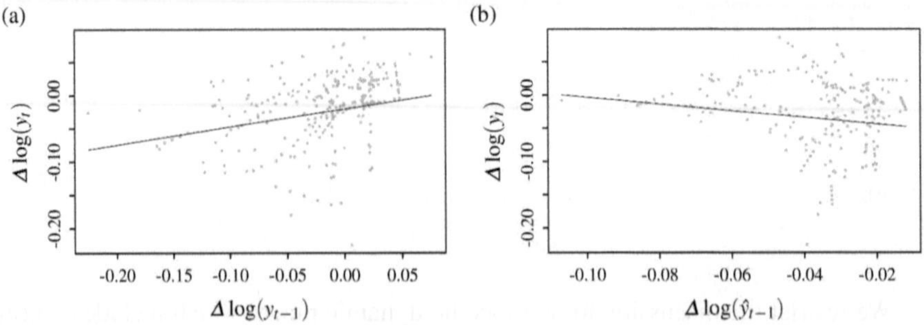

Fig. 4.6 Bivariate plots for the original and instrumented variable. (**a**) Original relation. (**b**) Relation for instrumented variable

type production function including lagged output as an additional regressor using again the Arellano and Bond data.

```
dly <- diff(log(E$y))
dly_1 <- lag(diff(log(E$y)),1)
dly_2 <- lag(diff(log(E$y)),2)
reg1 <- plm(lag(diff(log(y)),1)~lag(diff(log(y)),2),
             model="pooling",effect=NULL,data=E)
#summary(reg1)
r2 <- round(summary(reg1)$r.squared[1],3)*100;r2

## rsq
## 11.9

fit.dly_1 <- reg1$model[[1]]-residuals(reg1)
```

We observe that the instrument $\Delta y_{i,t-2}$ is rather weak as it explains only about 11.9 % of the variation of $\Delta y_{i,t-1}$. We show the simple linear association between $\Delta y_{i,t-1}$ and $\Delta y_{i,t}$ in the left panel and between instrumented $\Delta \hat{y}_{i,t-1}$ (fit.dly_1) and $\Delta y_{i,t}$ in the right panel (see Fig. 4.6).

```
nna <- !is.na(dly_1)
plot(dly_1,dly,pch=19,cex=0.5,col="darkgrey")
lines(dly_1[nna],fitted(lm(dly[nna]~dly_1[nna])))
nna <- !is.na(dly_2)
plot(fit.dly_1,dly[nna],pch=19,cex=0.5,col="darkgrey")
lines(fit.dly_1,fitted(lm(dly[nna]~fit.dly_1)))
```

Using the fitted values $\Delta \hat{y}_{i,t-1}$ as new regressor instead of $\Delta y_{i,t-1}$ we obtain the following results:

```
E2 <- pdata.frame(merge(E,cbind(fit.dly_1,index(fit.dly_1)),
                        all=T))
reg2 <- plm(diff(log(y))~fit.dly_1+diff(log(emp))+
              diff(log(cap)),
            model="pooling",effect=NULL,data=E2)
summary(reg2)
```

```
## Oneway (individual) effect Pooling Model
##
## Call:
## plm(formula = diff(log(y)) ~ fit.dly_1 + diff(log(emp)) + diff(log(cap)),
##     data = E2, effect = NULL, model = "pooling")
##
## Unbalanced Panel: n=140, T=4-6, N=611
##
## Residuals :
##      Min.   1st Qu.    Median   3rd Qu.      Max.
## -0.20400 -0.03320  0.00802  0.03700  0.13500
##
## Coefficients :
##                 Estimate Std. Error t-value    Pr(>|t|)
## (Intercept)     -0.05144    0.00484  -10.62     < 2e-16
## fit.dly_1       -0.60929    0.11875   -5.13 0.00000039
## diff(log(emp))   0.06806    0.01744    3.90    0.00011
## diff(log(cap))   0.02260    0.01640    1.38    0.16885
##
## (Intercept)     ***
## fit.dly_1       ***
## diff(log(emp)) ***
## diff(log(cap))
## ---
## Signif. codes:
## 0 '***' 0.001 '**' 0.01 '*' 0.05 '.' 0.1 ' ' 1
##
## Total Sum of Squares:    1.87
## Residual Sum of Squares: 1.73
## R-Squared          :  0.0748
##       Adj. R-Squared :  0.0743
## F-statistic: 16.3547 on 3 and 607 DF, p-value: 0.000000000312
```

Because of the weak instrument $\Delta y_{i,t-2}$ the estimate of the parameter of the lagged output $\hat{\rho}$ obtained through this instrumenting approach is obviously meaningless.

The instrumentation rests on the assumption that the production of a previous period is predetermined and hence not correlated with the current error term. According to this reasoning the number of available instruments increases with t:

period	regressor $y_{i,t-1} - y_{i,t-2}$	error term	instruments
$t = 3$	$y_{i,2} - y_{i,1}$	$v_{i3} - v_{i,2}$	y_1
$t = 4$	$y_{i,3} - y_{i,2}$	$v_{i4} - v_{i,3}$	y_1, y_2
$t = 5$	$y_{i,4} - y_{i,3}$	$v_{i5} - v_{i,4}$	y_1, y_2, y_3
$\vdots$	$\vdots$	$\vdots$	$\vdots$
$t = T$	$y_{i,T-1} - y_{i,T-2}$	$v_{iT} - v_{i,T-1}$	$y_1, y_2, \ldots, y_{T-2}$

This results in a large number of instruments $\sum_{t=1}^{T-2} t$, e.g., if $T = 10$ we may use 36 instruments. If the panel is unbalanced, the number of instruments will vary for different firms. This estimator is named General Method of Moments (GMM) estimator based on the orthogonality conditions (moments) of instruments and error terms (cf. Arellano and Bond (1991)) and implemented in the plm-package (the pgmm() function). If the error term is originally free of autocorrelation, then the error terms in differences will be negatively correlated, because Δv_{it}

contains $-v_i, t-1$ and $\Delta v_{i,t-1}$ contains $v_i, t-1$. This presumably present autocorrelation caused by differencing is considered in the estimation.

```
reg3 <- pgmm(log(y) ~ lag(log(y),1)+log(emp)+log(cap)
        | lag(log(y),2:10),
        effect="individual",
        model="onestep",data=E)
summary(reg3)

## Oneway (individual) effect One step model
##
## Call:
## pgmm(formula = log(y) ~ lag(log(y), 1) + log(emp) + log(cap) |
##     lag(log(y), 2:10), data = E, effect = "individual", model = "onestep")
##
## Unbalanced Panel: n=140, T=7-9, N=1031
##
## Number of Observations Used:  751
##
## Residuals
##      Min.   1st Qu.    Median      Mean   3rd Qu.      Max.
## -0.24100 -0.02060  0.00000 -0.00117  0.01750  0.17400
##
## Coefficients
##                 Estimate Std. Error z-value   Pr(>|z|)
## lag(log(y), 1)    0.7477     0.0491   15.22    < 2e-16
## log(emp)          0.1522     0.0296    5.14 0.00000027
## log(cap)         -0.0530     0.0156   -3.39    0.00069
##
## lag(log(y), 1) ***
## log(emp)       ***
## log(cap)       ***
## ---
## Signif. codes:
## 0 '***' 0.001 '**' 0.01 '*' 0.05 '.' 0.1 ' ' 1
##
## Sargan Test: chisq(27) = 131.2 (p.value=1.2e-15)
## Autocorrelation test (1): normal = -3.199 (p.value=0.00138)
## Autocorrelation test (2): normal = -4.912 (p.value=0.000000901)
## Wald test for coefficients: chisq(3) = 2300 (p.value=<2e-16)
```

Note that according to the test results given in the summary, there is autocorrelation present in the residuals as well as in the differenced residuals. Furthermore, the assumption that the instruments are valid (i.e., uncorrelated with the error term) seems also to be violated.

4.4 Recommended Readings

A very readable textbook on panel data econometrics is

- Hsiao C (2003) Analysis of panel data. Cambridge University Press, Cambridge.

An alternative textbook albeit less didactical is

- Baltagi BH (2013) Econometric analysis of panel data, 5th edn. Wiley, New York.

A brief overview provides the chapter on panel data in

- Greene WH (2011) Econometric analysis, 7th edn. Pearson Education, New Jersey.

The description of the plm-package also provides a very nice overview of panel data models and contains many exemplary analyses:

- Croissant Y, Millo G (2008) Panel data econometrics in R: The plm package. J Stat Software 27(2), URL http://www.jstatsoft.org/v27/i02/.

Many different aspects of estimating production functions are discussed in

- Griliches Z, Mairesse J (1995) Production functions: The search for identification. NBER Working Paper Series pp 169–203.

Two specific papers for instrumenting are

- Anderson T, Hsiao C (1981) Estimation of dynamic models with error components. J Am Stat Assoc 76:598–606
- Arellano M, Bond S (1991) Some tests of specification for panel data : Monte carlo evidence and an application to employment equations. Rev Econ Stud 58:277–297.

4.5 Exercises

The R-package "plm" (`library(plm)`) contains a variety of panel data methods and panel data sets. In this exercise, we will use some functions of plm and the data set "EmpUK" available in the package which has been studied by Arellano and Bond (1991).

1. The data structure
   a. Load the data into the work space using the command `data("EmplUK", package="plm")`.
   b. Try to get an overview of the data using `[]`, `dim()` and `summary()`.
   c. How many firms have been observed?
   d. Derive the distribution of spell length (that is: how many firms have been observed for seven years, eight years, ...).
   e. The package plm has a special type of data.frame suitable to handle the panel data structure by means of an index attribute. Read the information provided by the help page on the function `pdata.frame()`.
   f. Create a pdata.frame using the command `E <- pdata.frame(EmplUK)`.
   g. Draw a scatter plot with data of firm no. 77 only using the variables 'emp' (x) and 'output' (y).
   h. Draw a scatter plot with 'emp' (L) and 'output' (Y).
   i. Draw a scatter plot with 'capital' (K) and 'output' (Y).

j. Draw a scatter plot with 'ln(emp)' (l) and 'ln(output)' (y).
k. Draw a scatter plot with 'ln(capital)' (k) and 'ln(output)' (y).
l. Draw a scatter plot with firm-means, $\bar{l}_i$ and $\bar{y}_i$ using the command `between()`.
m. Draw a scatter plot with firm-means, $\bar{k}_i$ and $\bar{y}_i$ using the command `between()`.
n. Draw a scatter plot with deviations from firm specific means, $l_{it} - \bar{l}_i$ and $y_{it} - \bar{y}_i$ using the command `Within()`.
o. Draw a scatter plot with deviations from firm specific means, $k_{it} - \bar{k}_i$ and $y_{it} - \bar{y}_i$ using the command `Within()`.
p. Add the correlation coefficient to all your plots.

2. Consider a simple one-factor production function with firm specific intercepts and slopes

$$\ln(Y)_{it} = \beta_{0i} + \beta_{1i}\ln(L)_{it} + u_{it} \qquad \text{with } i = 1,\ldots,n;\ t = 1,\ldots,T.$$

a. Draw the bivariate plot.
b. Explain the difference between estimating this regression function and estimating n individual time series regressions for the n firms.
c. Estimate the regression function for all $n \cdot T$ data points using the function `lm()`.
d. Estimate the n regressions for individual firms and add the n regression lines to your scatter plot (hint: `$fitted` or `fitted()`). Be careful not to join lines from different firms.

3. Consider a simple pooled one-factor production function

$$\ln(Y)_{it} = \beta_0 + \beta_1\ln(L)_{it} + u_{it} \qquad \text{with } i = 1,\ldots,n;\ t = 1,\ldots,T.$$

a. Estimate the regression function using the commands `lm()` and `plm(..., model="pooling")` and compare your results.
b. Add the pooled regression line to your plot.

4. Consider a simple within one-factor production function

$$\ln(Y)_{it} = \beta_{0i} + \beta_1\ln(L)_{it} + u_{it} \qquad \text{with } i = 1,\ldots,n;\ t = 1,\ldots,T.$$

a. Draw the bivariate plot.
b. Estimate the regression function using the commands `lm()` and `plm(..., model="within")` and compare your results.
c. Add the regression lines to your plot. Be careful not to join lines from different firms.

5. Consider a simple random one-factor production function

$$\ln(Y)_{it} = \beta_0 + \beta_1\ln(L)_{it} + \alpha_i + v_{it} \qquad \text{with } i = 1,\ldots,n;\ t = 1,\ldots,T.$$

a. Estimate the slope parameter β_1 for the model containing a random firm-effect α_i using `plm(...,model="random")`.
b. Compare this parameter estimate with the estimates you obtain in the between and within models.
c. Add the between regression line to your (dummy-variable-) plot.

References

Anderson T, Hsiao C (1981) Estimation of dynamic models with error components. J Am Stat Assoc 76:598–606

Arellano M, Bond S (1991) Some tests of specification for panel data : Monte carlo evidence and an application to employment equations. Rev Econ Stud 58:277–297

Baltagi BH (2013) Econometric analysis of panel data, 5th edn. Wiley, New York

Behr A (2003) A comparison of dynamic panel data estimators: Monte carlo evidence and an application to the investment function. Discussion paper of the Economic Research Centre of the Deutsche Bundesbank 03(5)

Croissant Y, Millo G (2008) Panel data econometrics in R: The plm package. J Stat Software 27(2), URL http://www.jstatsoft.org/v27/i02/

Greene WH (2011) Econometric analysis, 7th edn. Pearson Education, New Jersey

Griliches Z, Mairesse J (1995) Production functions: The search for identification. NBER Working Paper Series pp 169–203

Hsiao C (2003) Analysis of panel data. Cambridge University Press, Cambridge

Linear Programming 5

Abstract

In production economics, we often want to maximize (e.g., output) or minimize (e.g., costs) an objective function under inequality constraints. When restricting the problem to be linear both in the objective function and in the inequality constraints, the simplex-tableau is a convenient way to formulate the problem and to obtain the solution using the simplex method. R provides several functions to obtain the solution of linear programming problems which will be used to solve some exemplary problems.

5.1 Introduction

In economics, we often face optimization problems. For example we want to maximize the likelihood function for a parameter vector given the observations. These problems are typically nonlinear. If we face constraints, these are typically equality constraints. In linear programming, we face problems that have a linear objective function and inequality constraints. For example the quantity of an input that can be used is limited by the amount in stock. Or there might be some capacity constraints which have to be met in the production process. In general, we deal with quantities in production economics that must be non-negative.

5.1.1 Cost Minimization

We start with an example:[1] Two kinds of food (inputs) contain amounts of three different minerals. There are two inputs: x_1 and x_2. The input prices are $w_1 = 0.6$

[1]We follow closely Chiang (1984), chapters 19 and 20.

A. Behr, *Production and Efficiency Analysis with R*,
DOI 10.1007/978-3-319-20502-1_5

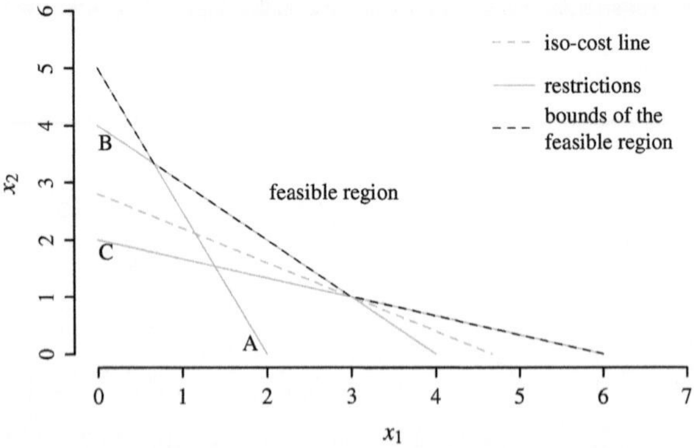

Fig. 5.1 Cost minimization example: x_1/x_2-diagram

for input 1 and $w_2 = 1$ for input 2. There are three outputs y_1, y_2, and y_3. Each input "produces" outputs in a fixed ratio of quantities. To meet the demand for the goods, there must be certain quantities of outputs produced: 20 of good 1, 20 of good 2, and 12 of good 3. Our aim is to produce at least the required amounts of the output goods spending as little as possible for the inputs.

We formulate the problem as a cost minimization problem:

$$\begin{aligned} &\text{objective function } C = 0.6x_1 + x_2 \\ &\text{restrictions} \qquad \text{A: } y_1 : 10x_1 + 4x_2 \geq 20 \\ &\qquad\qquad\qquad\quad \text{B: } y_2 : 5x_1 + 5x_2 \geq 20 \\ &\qquad\qquad\qquad\quad \text{C: } y_3 : 2x_1 + 6x_2 \geq 12 \\ &\text{non-negativity} \quad x_1, x_2 \geq 0. \end{aligned} \tag{5.1}$$

We display the problem using a x_1, x_2-diagram (see Fig. 5.1). The objective function (linear cost function) is a negative sloping line

$$C = 0.6x_1 + x_2 \quad \Longrightarrow \quad x_2 = C - 0.6x_1. \tag{5.2}$$

The closer this objective function (iso-cost line) is located to the origin, the lower the costs. The three inequality constraints (here expressed as equalities for the purpose of drawing) are also downward sloping:

$$\begin{aligned} y_1 : 10x_1 + 4x_2 = 20 \quad &\Longrightarrow \quad x_2 = 5 - \frac{5}{2}x_1 \\ y_2 : 5x_1 + 5x_2 = 20 \quad &\Longrightarrow \quad x_2 = 4 - x_1 \\ y_3 : 2x_1 + 6x_2 = 12 \quad &\Longrightarrow \quad x_2 = 2 - \frac{1}{3}x_1. \end{aligned} \tag{5.3}$$

Only x_1, x_2 combinations to the north-east of the inequality constraints are feasible. Other combinations violate the constraints as the output falls below the required level. Hence, our aim is to move the objective function as close to the origin as possible without violating the inequality constraints, that is without leaving the feasible region.

We observe that the lines for inequality restrictions cross each other. The constraints for y_1 and y_2 cross at

$$\begin{aligned} & y_1 = y_2 : \\ & x_2 = 5 - \frac{5}{2}x_1 = 4 - x_1 \quad \Longrightarrow \quad x_1 = \frac{2}{3}; x_2 = \frac{10}{3}. \end{aligned} \tag{5.4}$$

The constraints for y_2 and y_3 cross at

$$\begin{aligned} & y_2 = y_3 : \\ & x_2 = 4 - x_1 = 2 - \frac{1}{3}x_1 \quad \Longrightarrow \quad x_1 = 3; x_2 = 1. \end{aligned} \tag{5.5}$$

Additionally, we also obtain the point where restriction of y_1 crosses the x_2-axis

$$\begin{aligned} & y_1 : \\ & x_2 = 5 - \frac{5}{2}x_1 = 5 - \frac{5}{2} \cdot 0 = 5 \end{aligned} \tag{5.6}$$

and the points where restriction of y_3 crosses the x_1-axis

$$\begin{aligned} & y_3 : \\ & x_2 = 2 - \frac{1}{3}x_1 \quad \Longrightarrow \quad x_1 = 6 - x_2 = 6 - 0 = 6. \end{aligned} \tag{5.7}$$

Note that (with the rare exception that a restriction has the same slope as the objective function) the cost minimizing point will be one of the extreme points of the feasible region. Therefore, the strategy is to walk through these extreme points and compare the resulting costs. With the known linear cost function (see Eq. (5.2))

$$C(x_1, x_2) = 0.6x_1 + x_2 \tag{5.8}$$

we obtain the costs at the four extreme points as

$$\begin{aligned} C(0, 5) &= 0.6 \cdot 0 + 5 = 5 \\ C(2/3, 10/3) &= 0.6 \cdot 2/3 + 10/3 = 3.7333 \\ C(3, 1) &= 0.6 \cdot 3 + 1 = 2.8 \\ C(6, 0) &= 0.6 \cdot 6 + 0 = 3.6. \end{aligned} \tag{5.9}$$

The objective function is closest to the origin in point (3, 1) with costs 2.8. Given these costs, we can find the points at which the iso-cost line (see Eq. (5.2)) crosses the axes:

$$\begin{aligned} x_1 &= \frac{1}{0.6}(C - x_2) = \frac{1}{0.6}(2.8 - 0) = 4.6667 \\ x_2 &= C - 0.6x_1 = 2.8 - 0.6 \cdot 0 = 2.8. \end{aligned} \tag{5.10}$$

5.1.2 Profit Maximization

Now we look at a maximization problem. Assume we have two outputs x_1 and x_2 and output prices are $c_1 = 40$ for x_1 and $c_2 = 30$ for x_2. There are three production restrictions (A, B, C).

The problem set up is

$$\begin{aligned} &\text{objective function } \pi = 40x_1 + 30x_2 \\ &\text{restrictions} \quad A: x_1 \leq 16 \\ &\qquad\qquad\quad B: x_2 \leq 8 \\ &\qquad\qquad\quad C: x_1 + 2x_2 \leq 24 \\ &\text{non-negativity} \quad x_1, x_2 \geq 0 \end{aligned} \tag{5.11}$$

We illustrate the problem using a x_1, x_2-diagram (see Fig. 5.2). The objective function is

$$\pi = 40x_1 + 30x_2 \implies x_2 = \frac{\pi}{30} - \frac{4}{3}x_1. \tag{5.12}$$

The profit increases with the amount of x_1 and x_2. Hence, we aim to move the iso-profit line as far to the north-east as possible without violating the restrictions. We

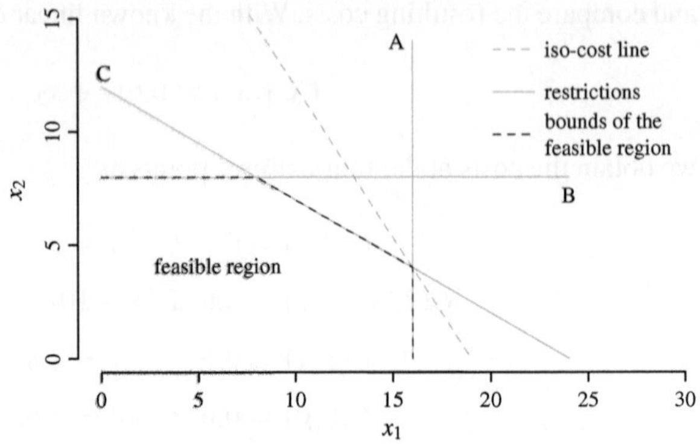

Fig. 5.2 Profit maximization example: x_1/x_2-diagram

insert the three lines representing the inequality constraints:

$$\begin{aligned} &A : x_1 = 16 \\ &B : x_2 = 8 \\ &C : x_1 + 2x_2 = 24 \quad \Longrightarrow \quad x_2 = 12 - \frac{1}{2}x_1. \end{aligned} \tag{5.13}$$

We obtain the point where restrictions A and C cross

$$\begin{aligned} &A = C : \\ &x_1 = 16 \quad \Longrightarrow \quad x_2 = 12 - \frac{1}{2}x_1 = 12 - 8 = 4 \end{aligned} \tag{5.14}$$

and where restrictions B and C cross

$$\begin{aligned} &B = C : \\ &x_2 = 8 \quad \Longrightarrow \quad x_1 = 24 - 2x_2 = 24 - 16 = 8. \end{aligned} \tag{5.15}$$

Additionally, the feasible region has the two extreme points where restriction B meets the x_2-axis $(0, 8)$ and where restriction A meets the x_1-axis $(16, 0)$. To find the maximum, we compare the profit at the extreme points of the feasible region

$$\begin{aligned} \pi(x_1, x_2) &= 40 \cdot x_1 + 30 \cdot x_2 \\ \pi(0, 8) &= 40 \cdot 0 + 30 \cdot 8 = 240 \\ \pi(8, 8) &= 40 \cdot 8 + 30 \cdot 8 = 560 \\ \pi(16, 4) &= 40 \cdot 16 + 30 \cdot 4 = 760 \\ \pi(16, 0) &= 40 \cdot 16 + 30 \cdot 0 = 640. \end{aligned} \tag{5.16}$$

Choosing the profit maximizing point $(16, 4)$ with profit 760

$$40 \cdot x_1 + 30 \cdot x_2 = 760, \tag{5.17}$$

we find the points where the iso-profit line meets the axes

$$x_2 = \frac{760}{30} = 25.333 \quad \text{and} \quad x_1 = \frac{760}{40} = 19. \tag{5.18}$$

5.1.3 General Notation

In the following, we state the maximization and the minimization problems in a very general way. We also show the shorthand exposition using matrix notation.

Maximization problem:

$$\begin{array}{lll} \text{objective function (max)} & \pi = \sum_{j=1}^{n} c_j x_j & \\ \text{restrictions} & \sum_{j=1}^{n} a_{ij} x_j \leq r_i \quad (i = 1, \ldots, m) & \\ \text{non-negativity} & x_j \geq 0 \quad (j = 1, \ldots, n) & \end{array} \tag{5.19}$$

In matrix notation:

$$\begin{array}{ll} \text{objective function (max)} & \pi = \underset{(1\times n)}{c'} \underset{(n\times 1)}{x} \\ \text{restrictions} & \underset{(m\times n)}{A} \underset{(n\times 1)}{x} \leq \underset{(m\times 1)}{r} \\ \text{non-negativity} & \underset{(n\times 1)}{x} \geq \underset{(n\times 1)}{0} \end{array} \tag{5.20}$$

Minimization problem:

$$\begin{array}{ll} \text{objective function (min)} & C = \sum_{j=1}^{n} c_j x_j \\ \text{restrictions} & \sum_{j=1}^{n} a_{ij} x_j \geq r_i \quad (i = 1, \ldots, m) \\ \text{non-negativity} & x_j \geq 0 \quad (j = 1, \ldots, n) \end{array} \tag{5.21}$$

In matrix notation:

$$\begin{array}{ll} \text{objective function (min)} & C = \underset{(1\times n)}{c'} \underset{(n\times 1)}{x} \\ \text{restrictions} & \underset{(m\times n)}{A} \underset{(n\times 1)}{x} \geq \underset{(m\times 1)}{r} \\ \text{non-negativity} & \underset{(n\times 1)}{x} \geq \underset{(n\times 1)}{0} \end{array} \tag{5.22}$$

5.1.4 Convex Sets

For $n > 2$, variables minimization and maximization problems cannot be solved graphically. We have to apply numerical rather than graphical methods. Still, only the extreme points of the convex feasible set have to be examined. The strategy is to move along the extreme points to compare the value of the objective function. A convex set S has the property that each point on the linear connection between two points belonging to set S will also belong to set S:

$$\left.\begin{array}{l} u \in S \\ v \in S \end{array}\right\} \Longrightarrow w \in S \quad \text{where} \quad w = \theta u + (1-\theta) v \quad \text{with} \quad (0 \leq \theta \leq 1). \tag{5.23}$$

5.1.5 Slacks and Surpluses

For applying numerical methods, the use of slack variables is helpful. With the help of slack variables, the inequality constraints can be expressed as equalities. In the solution usually not all of the inequality constraints will be met exactly. Some constraints will not be binding, hence there is some slack (excess capacity, underutilization) of inputs or some surplus (overproduction) of outputs.

Using a slack variable $s \geq 0$, the inequality restriction $x \leq 16$ can be stated as an equality $x + s = 16$. Because s is non-negative, the inequality restriction will not be violated. If $s > 0$ the inequality constraint is not binding. Using three slack variables s_1, s_2, and s_3, the profit maximization problem with two outputs and three production restrictions can be stated as follows:

$$\begin{array}{ll} \text{objective function (max)} & \pi = 40x_1 + 30x_2 + 0s_1 + 0s_2 + 0s_3 \\ \text{restrictions} & A: x_1 + s_1 = 16 \\ & B: x_2 + s_2 = 8 \\ & C: x_1 + 2x_2 + s_3 = 24 \\ \text{non-negativity} & x_1, x_2, s_1, s_2, s_3 \geq 0 \end{array} \tag{5.24}$$

In matrix notation, we note the problem as

$$\begin{bmatrix} 1\,0\,1\,0\,0 \\ 0\,1\,0\,1\,0 \\ 1\,2\,0\,0\,1 \end{bmatrix} \begin{bmatrix} x_1 \\ x_2 \\ s_1 \\ s_2 \\ s_3 \end{bmatrix} = \begin{bmatrix} 16 \\ 8 \\ 24 \end{bmatrix}. \tag{5.25}$$

Comparing again the five possible solutions (extreme values), we find the following values of the slack variables

output (x_1, x_2)	solution $(x_1, x_2, s_1, s_2, s_3)$	target value π
$(0, 0)$	$(0, 0, 16, 8, 24)$	0
$(16, 0)$	$(16, 0, 0, 8, 8)$	640
$(16, 4)$	$(16, 4, 0, 4, 0)$	760
$(8, 8)$	$(8, 8, 8, 0, 0)$	560
$(0, 8)$	$(0, 8, 16, 0, 8)$	240

We observe that the solution with a profit of 760 has output quantities $x_1 = 16, x_2 = 4$ and $s_1 = 0, s_2 = 4, s_3 = 0$. Therefore, the restriction B is not binding and there is a slack $s_2 = 4$.

5.2 Simplex Algorithm

The basic idea of the simplex algorithm is moving from one extreme point to the next, each time calculating the value of the maximand (minimand) until the optimal extreme point is found.

5.2.1 Maximization

We combine the objective function

$$\begin{aligned}\pi &= 40x_1 + 30x_2 \\ 1\pi - 40x_1 - 30x_2 &= 0\end{aligned} \tag{5.26}$$

and the rows of the constraints including the slack variables which are used to transform the inequality constraints into equality constraints. The simplex tableau is given by

π	x_1	x_2	s_1	s_2	s_3	const.
1	−40	−30	0	0	0	0
0	1	0	1	0	0	16
0	0	1	0	1	0	8
0	1	2	0	0	1	24

One (trivial) solution (S_1) is the origin $x_1 = 0, x_2 = 0$

$$S_1 : \pi = x_1 = x_2 = 0, s_1 = 16, s_2 = 8, s_3 = 24. \tag{5.27}$$

Of course, $\pi = 0$ is no optimal solution. Without any production, the slacks are obtained as the constants of the inequality restrictions. Note that the objective function and the equality constraints are linearly independent:

π	s_1	s_2	s_3	const.
1	0	0	0	0
0	1	0	0	16
0	0	1	0	8
0	0	0	1	24

An additional unit of x_1 increases the profit by $c_1 = 40$ (compared to $c_2 = 30$). Therefore, we choose x_1 as the pivot-column.

π	$[x_1]$	x_2	s_1	s_2	s_3	const.
1	−40	−30	0	0	0	0
0	1	0	1	0	0	16
0	0	1	0	1	0	8
0	1	2	0	0	1	24

To obtain linear independent columns, the pivot-column (x_1) has to be transformed into the vector $(0, 1, 0, 0)$. We choose the pivot-element with the smallest ratio of slack to the quantity (replacement ratio) of x_1 (here: $16/1 < 24/1$.) The pivot-element [1] (at position 2, 2) is indicated by brackets.

π	x_1	x_2	s_1	s_2	s_3	const.
1	−40	−30	0	0	0	0
0	[1]	0	1	0	0	16
0	0	1	0	1	0	8
0	1	2	0	0	1	24

Next, we multiply row 2 by 40 and add the result to row 1. We then subtract row 2 from row 4. This results in:

π	x_1	x_2	s_1	s_2	s_3	const.
1	0	−30	40	0	0	640
0	[1]	0	1	0	0	16
0	0	1	0	1	0	8
0	0	2	−1	0	1	8

Ignoring columns x_2 and s_1, we obtain the following tableau:

π	x_1	s_2	s_3	const.
1	0	0	0	640
0	[1]	0	0	16
0	0	1	0	8
0	0	0	1	8

We now have obtained solution S_2 with a profit of 640.

$$S_2 : \pi = 640; x_1 = 16; x_2 = 0, s_1 = 0, s_2 = 8, s_3 = 8. \tag{5.28}$$

Next we can increase the amount of x_2 choosing [1] (position 4, 3) as the pivot-element after dividing row 4 by 2.

π	x_1	x_2	s_1	s_2	s_3	const.
1	0	−30	40	0	0	640
0	1	0	1	0	0	16
0	0	1	0	1	0	8
0	0	[1]	−0.5	0	0.5	4

We convert this column into the vector (0, 0, 0, 1) by multiplying row 4 by 30, adding the result to row 1 and subtracting row 4 from row 3.

π	x_1	x_2	s_1	s_2	s_3	const.
[1]	0	0	25	0	15	760
0	[1]	0	1	0	0	16
0	0	0	0.5	[1]	−0.5	4
0	0	[1]	−0.5	0	0.5	4

In each of the columns π, x_1, x_2, s_2, we now have obtained only zeros and a single 1. The constants indicate the solution:

$$S_3 : \pi = 760; x_1 = 16; x_2 = 4, s_1 = 0, s_2 = 4, s_3 = 0. \tag{5.29}$$

Graphically, we find that we started from the origin (S_1), walked through S_2 to finally end up at S_3.

5.2.2 Minimization

The minimization problem

$$\begin{aligned} &\text{objective function (max)} && C = 12y_1 + 42y_2 \\ &\text{restrictions} && \begin{bmatrix} 1 & 2 \\ 1 & 4 \\ 3 & 1 \end{bmatrix} \begin{bmatrix} y_1 \\ y_2 \end{bmatrix} \geq \begin{bmatrix} 3 \\ 4 \\ 3 \end{bmatrix} \\ &\text{non-negativity} && y_1, y_2 \geq 0 \end{aligned} \tag{5.30}$$

We add the slack variables s_1, s_2, s_3 to transform the inequality constraints into equalities. Because $s_i \geq 0$, we insert a negative identity matrix

C	y_1	y_2	s_1	s_2	s_3	const.
1	−12	−42	0	0	0	0
0	1	2	−1	0	0	3
0	1	4	0	−1	0	4
0	3	1	0	0	−1	3

We try to start in the origin $y_1 = y_2 = 0$ but obtain the values $\{-3, -4, -3\}$ for the slack variables s_1, s_2, s_3. Because of the non-negativity constraint $s_i \geq 0$, the origin is not feasible. Therefore, we have to find another extreme point to start our search with. To make sure that the non-negativity constraints are met, we introduce further auxiliary variables v_i with prohibitive costs (e.g., -100). These prohibitive costs ensure that these artificial variables will enter the solution with values $v_1 = v_2 = v_3 = 0$.

C	y_1	y_2	s_1	s_2	s_3	v_1	v_2	v_3	const.
1	−12	−42	0	0	0	−100	−100	−100	0
0	1	2	−1	0	0	1	0	0	3
0	1	4	0	−1	0	0	1	0	4
0	3	1	0	0	−1	0	0	1	3

To set the variables v_i in the objective row to 0, we add each of the rows 2, 3, 4 multiplied by 100 to row 1. We obtain the following tableau 1, our initial basic feasible solution. Now we choose the pivot-element. We look for the column with the highest cost (here 658) per unit of output and the row according to the lowest replacement ratio ($4/4 < 3/2 < 3/1$).

C	y_1	y_2	s_1	s_2	s_3	v_1	v_2	v_3	const.
1	488	658	−100	−100	−100	0	0	0	1000
0	1	2	−1	0	0	1	0	0	3
0	1	[4]	0	−1	0	0	1	0	4
0	3	1	0	0	−1	0	0	1	3

In tableau 1 we divide row 3 by 4. Then we subtract the new row 3 (pivot-element now is 1) from row 4, subtract two times row 3 from row 2 and finally we subtract 658 times row 3 from row 1. This results in tableau 2, where we choose a new pivot-element (indicated by brackets). Because of the high value of y_1, we choose column

2 and the fourth row is chosen because it has the lowest replacement ratio.

C	y_1	y_2	s_1	s_2	s_3	v_1	v_2	v_3	const.
1	323.5	0	−100	64.5	−100	0	−164.5	0	342
0	0.5	0	−1	0.5	0	1	−0.5	0	1
0	0.25	1	0	−0.25	0	0	0.25	0	1
0	[2.75]	0	0	0.25	−1	0	−0.25	1	2

We then divide row 4 by 2.75 to transform the pivot-element to 1. Next, we subtract a quarter of the obtained row 4 from row 3, subtract a half of row 4 from row 2, and finally subtract 323.5 times row 4 from row 1. This results in tableau 3.

C	y_1	y_2	s_1	s_2	s_3	v_1	v_2	v_3	const.
1	0	0	−100	35.09	17.63	0	−135.09	−117.64	106.73
0	0	0	−1	[0.45]	0.18	1	−0.45	−0.18	0.64
0	0	1	0	−0.27	0.09	0	0.27	−0.09	0.82
0	1	0	0	0.09	−0.36	0	−0.09	0.36	0.73

We proceed in the same manner as before. We choose a pivot-element (row 2, column 5), transform this row to obtain the value 1 for the pivot-element and transform the remaining rows to obtain zeros in the column of the pivot-element. This results in tableau 4.

C	y_1	y_2	s_1	s_2	s_3	v_1	v_2	v_3	const.
1	0	0	−22.8	0	3.6	−77.20	−100	−103.6	57.6
0	0	0	−2.2	1	[0.4]	2.2	−1	−0.4	1.4
0	0	1	−0.6	0	0.2	0.6	0	−0.2	1.2
0	1	0	0.2	0	−0.4	−0.2	0	0.4	0.6

Choosing a pivot-element (row 2, column 6) and again doing some manipulations, we finally arrive at tableau 5 which contains the results of the optimization procedure.

C	y_1	y_2	s_1	s_2	s_3	v_1	v_2	v_3	const.
1	0	0	−3	−9	0	−97	−91	−100	[45]
0	0	0	−5.5	2.5	1	5.5	−2.5	−1	3.5
0	0	1	0.5	−0.5	0	−0.5	0.5	0	0.5
0	1	0	−2	1	0	2	−1	0	2

In this tableau, we find the following results: Minimal costs are $\bar{C} = 45$ (first row). Optimal quantities are $y_1 = 2$ and $y_2 = 0.5$. The third restriction is not binding with (slack) $s_3 = 3.5$. Note that the column for s_3 contains only zeros and a single 1 which indicates the row of the constant giving the amount of the slack. If we insert the optimal quantities of y_1 and y_2, we find the following results for the three restrictions:

$$\begin{aligned} A &: 2 + 1 = 3 \\ B &: 2 + 2 = 4 \\ C &: 6 + 0.5 \geq 3 \end{aligned} \tag{5.31}$$

The restrictions A and B are exactly met (no slacks) with slack variables $s_1 = s_2 = 0$. From restriction C, we obtain the slack $s_3 = 3.5$.

5.2.3 Duality

So far, we analyzed minimization and maximization problems differently. In fact, each minimization problem can be formulated as a maximization problem and vice versa. The original optimization problem is called the primal, the obtained problem the corresponding dual. Variables which have to be chosen are denoted by x_j in the primal and by y_i in the corresponding dual.

Example 1

$$\begin{array}{ll} \text{Primal} & \\ \text{objective function (max)} & \pi = 3x_1 + 4x_2 + 3x_3 \\ \text{restrictions} & \begin{bmatrix} 1 & 1 & 3 \\ 2 & 4 & 1 \end{bmatrix} \begin{bmatrix} x_1 \\ x_2 \\ x_3 \end{bmatrix} \leq \begin{bmatrix} 12 \\ 42 \end{bmatrix} \\ \text{non-negativity} & x_1, x_2, x_3 \geq 0 \end{array} \tag{5.32a}$$

$$\begin{array}{ll} \text{Dual} & \\ \text{objective function (min)} & \pi^* = 12y_1 + 42y_2 \\ \text{restrictions} & \begin{bmatrix} 1 & 2 \\ 1 & 4 \\ 3 & 1 \end{bmatrix} \begin{bmatrix} y_1 \\ y_2 \end{bmatrix} \geq \begin{bmatrix} 3 \\ 4 \\ 3 \end{bmatrix} \\ \text{non-negativity} & y_1, y_2 \geq 0 \end{array} \tag{5.32b}$$

Example 2

$$\begin{array}{ll} \text{Primal} & \\ \text{objective function (min)} & C = 4x_1 + 3x_2 + 8x_3 \\ \text{restrictions} & \begin{bmatrix} 1 & 0 & 1 \\ 0 & 1 & 2 \end{bmatrix} \begin{bmatrix} x_1 \\ x_2 \\ x_3 \end{bmatrix} \geq \begin{bmatrix} 2 \\ 5 \end{bmatrix} \\ \text{non-negativity} & x_1, x_2, x_3 \geq 0 \end{array} \tag{5.33a}$$

Dual

$$\begin{array}{ll} \text{objective function (max)} & C^* = 2y_1 + 5y_2 \\ \text{restrictions} & \begin{bmatrix} 1 & 0 \\ 0 & 1 \\ 1 & 2 \end{bmatrix} \begin{bmatrix} y_1 \\ y_2 \end{bmatrix} \leq \begin{bmatrix} 4 \\ 3 \\ 8 \end{bmatrix} \\ \text{non-negativity} & y_1, y_2 \geq 0 \end{array} \tag{5.33b}$$

The general transformation of the primal into the dual is given in matrix notation:

Example 1

$$\begin{array}{lll} \text{Primal} & & \text{Dual} \\ \text{obj. funct. (max) } \pi = c'x & \longrightarrow & \text{obj. funct. (min) } \pi^* = r'y \\ \text{restrictions} \quad Ax \leq r & & \text{restrictions} \quad A'y \geq c \\ \text{non-negativity} \quad x \geq 0 & & \text{non-negativity} \quad y \geq 0 \end{array} \tag{5.34}$$

Example 2

$$\begin{array}{lll} \text{Primal} & & \text{Dual} \\ \text{obj. funct. (min) } C = c'x & \longrightarrow & \text{obj. funct. (max) } C^* = r'y \\ \text{restrictions} \quad Ax \geq r & & \text{restrictions} \quad A'y \leq c \\ \text{non-negativity} \quad x \geq 0 & & \text{non-negativity} \quad y \geq 0 \end{array} \tag{5.35}$$

Duality allows to choose the representation of the problem which can be solved more easily. If the primal has m restrictions and n variables, the matrix A has dimension $m \times n$. The corresponding dual accordingly has n restrictions and m variables. For the examples above (Eq. 5.32a, 5.33a), the dual can still be solved graphically because it has 2 variables (the primal has 3 variables). In general, for applying the simplex algorithm, we like to choose the formulation with less restrictions. And because maximization is most often easier than minimization, we often like to choose the dual of a minimization problem. In the following, solutions are indicated with a bar. For example $\bar{y}_i$ is the solution of the dual $\bar{C}^*$ and $\bar{x}_j$ is the solution of the primal $\bar{C}$. s denotes the slack variables in the primal and t denotes the slack variables in the dual.

The close correspondence between the primal and the dual can be further utilized as the two theorems state:

Duality Theorem I: The optimal values of the primal and the dual objective functions are always identical, provided that optimal feasible solutions do exist; that is $C = C^*$ and $\pi = \pi^*$.

Duality Theorem II (a): If a certain choice variable in a linear program is optimally nonzero, then the corresponding slack variable in the counterpart program must be optimally zero. That is (using s_i to denote the ith primal slack variable, and t_j to denote the jth dual slack variable),

$$\bar{y}_i > 0 \Longrightarrow \bar{s}_i = 0 \quad \text{and} \quad \bar{x}_j > 0 \Longrightarrow \bar{t}_j = 0. \tag{5.36}$$

Duality Theorem II (b): If a certain slack variable in a linear program is optimally nonzero, then the corresponding choice variable in the counterpart program must be optimally zero. That is,

$$\bar{s}_i > 0 \Longrightarrow \bar{y}_i = 0 \quad \text{and} \quad \bar{t}_j > 0 \Longrightarrow \bar{x}_j = 0. \tag{5.37}$$

5.2.4 Simplex Tableau

We now apply the duality theorems to solve the following linear programming problem:

$$\begin{array}{ll} \text{Dual} & \\ \text{objective function (min)} & \pi^* = 12y_1 + 42y_2 \\ \text{restrictions} & \begin{bmatrix} 1 & 2 \\ 1 & 4 \\ 3 & 1 \end{bmatrix} \begin{bmatrix} y_1 \\ y_2 \end{bmatrix} \geq \begin{bmatrix} 3 \\ 4 \\ 3 \end{bmatrix} \\ \text{non-negativity} & y_1, y_2 \geq 0 \end{array} \tag{5.38}$$

We set up the simplex tableau in several steps. First, we define the first row containing the objective function, zero coefficients for the slack variables (t_1, t_2, t_3), and prohibitive costs (here 100, the negative sign results from setting the cost function to 0) for the artificial variables v_1, v_2, v_3 to help finding an initial solution to start with

$$\begin{aligned} \pi^* &= 12y_1 + 42y_2 \\ 1\pi^* - 12y_1 - 42y_2 &= 0. \end{aligned} \tag{5.39}$$

Because the restrictions are of the $\geq$-form, we subtract the non-negative slack variables ($s \geq 0$). Each restriction is supplemented with an artificial variable. This results in the following simplex tableau:

π^*	y_1	y_2	t_1	t_2	t_3	v_1	v_2	v_3	const.
1	−12	−42	0	0	0	−100	−100	−100	0
0	1	2	−1	0	0	1	0	0	3
0	1	4	0	−1	0	0	1	0	4
0	3	1	0	0	−1	0	0	1	3

The solution is given as:

C	y_1	y_2	t_1	t_2	t_3	v_1	v_2	v_3	const.
{1}	0	0	[−3]	[−9]	[0]	−97	−91	−100	[{45}]
0	0	0	−5.5	2.5	{1}	5.5	−2.5	−1	{3.5}
0	0	{1}	0.5	−0.5	0	−0.5	0.5	0	{0.5}
0	{1}	0	−2	1	0	2	−1	0	{2}

The solutions are indicated by brackets {} in the tableau:

$$\{\bar{\pi}^*, \bar{y}_1, \bar{y}_2, \bar{t}_1, \bar{t}_2, \bar{t}_3,\} = \{45, 2, 0.5, 0, 0, 3.5\}. \tag{5.40}$$

The corresponding primal of the dual

$$\begin{aligned}
&\text{Primal} \\
&\text{objective function (max) } \pi = 3x_1 + 4x_2 + 3x_3 \\
&\text{restrictions} \qquad \begin{bmatrix} 1 & 1 & 3 \\ 2 & 4 & 1 \end{bmatrix} \begin{bmatrix} x_1 \\ x_2 \\ x_3 \end{bmatrix} \leq \begin{bmatrix} 12 \\ 42 \end{bmatrix} \\
&\text{non-negativity} \qquad x_1, x_2, x_3 \geq 0
\end{aligned} \tag{5.41}$$

is also given in the solution-tableau. Solutions of the dual are indicated by brackets []:

$$\bar{\pi} = 45, \bar{x}_1 = 3, \bar{x}_2 = 9, \bar{x}_3 = 0, \bar{s}_1 = 0, \bar{s}_2 = 0. \tag{5.42}$$

According to duality theorem 2, we know that two (s_1 and s_2) out of the three slack variables will be 0 and that because the solution-value of the third slack variable is positive, x_3 will be 0 in the solution of the primal:

$$\begin{aligned}
\bar{y}_1 > 0; \bar{y}_2 > 0 &\Longrightarrow \bar{s}_1 = 0; \bar{s}_2 = 0 \\
\bar{t}_3 > 0 &\Longrightarrow \bar{x}_3 = 0.
\end{aligned} \tag{5.43}$$

Because of $\bar{x}_3 = 0$, the restrictions simplify towards

$$\begin{aligned}
x_1 + x_2 &= 12 \\
2x_1 + 4x_2 &= 42 \\
x_2 = 12 - x_1 &= 10.5 - 0.5x_1 \\
\Rightarrow \bar{x}_1 &= 3; \bar{x}_2 = 9.
\end{aligned} \tag{5.44}$$

5.3 Solving the Simplex Using R

In the preceding sections, we solved the linear programming problems either graphically (if the minimand (cost function) or maximand (profit function) had only up to two variables) or through setting up the simplex tableau and shuffling around rows to obtain the solution. Of course, linear programming problems can be solved very easily using the R programming environment.

5.3.1 The Setup of Simplex Tableaus

We have already introduced the shorthand notation for the profit maximization problem and its dual:

$$\begin{array}{llcll} \text{Primal} & & & \text{Dual} & \\ \text{obj. funct. (max)} & \pi = c'x & \longrightarrow & \text{obj. funct. (min)} & \pi^* = r'y \\ \text{restrictions} & Ax \leq r & & \text{restrictions} & A'y \geq c \\ \text{non-negativity} & x \geq 0 & & \text{non-negativity} & y \geq 0 \end{array} \tag{5.45}$$

In R, there are many different packages for solving linear programs. An easy to use function is `simplex()` which is included in the boot-package. This function will optimize the linear function $a \times x$, subject to the constraints $A1 \times x \leq b1, A2 \times x \geq b2, A3 \times x = b3$, and $x \geq 0$. The vector a of length n contains the coefficients of the objective function. The matrix $A1$ has dimension $m1 \times n$ and contains the coefficients for the $\leq$ type of constraints. $b1$ is a vector of length $m1$ giving the right-hand side of the $\leq$ constraints. All values in $b1$ must be non-negative. $A2$ is the $m2 \times n$ matrix of coefficients for the $\geq$ type of constraints. $b2$ is a vector of length $m2$ giving the right-hand side of the $\geq$ constraints. As in $b1$, all values in $b2$ must be non-negative. And finally, the matrix $A3$ is a $m3 \times n$ matrix of coefficients for the equality constraints. Here $b3$ is a vector of length $m3$ giving the right-hand side of equality constraints. By default, `simplex()` minimizes the linear function. The option `maxi=T` can be set for maximization problems.

5.3.2 Solving Some Examples

We first use `simplex()` to solve the following profit maximization problem which we already solved graphically above.

$$\begin{array}{ll} \text{objective function} & \pi = 40x_1 + 30x_2 \\ \text{restrictions} & A: x_1 \leq 16 \\ & B: x_2 \leq 8 \\ & C: x_1 + 2x_2 \leq 24 \\ \text{non-negativity} & x_1, x_2 \geq 0 \end{array} \tag{5.46}$$

We have vector a of length $n = 2$ and three inequality constraints of the $\leq$ type. Therefore, we have to define a 3×2 matrix $A1$ and a vector $b1$ of length 3.

```
library(boot)
simplex(a = c(40,30),
        A1 = rbind(c(1,0),c(0,1),c(1,2)),
        b1 = c(16,8,24),
        A2 = rbind(c(1,0),c(0,1)),
        b2 = c(0,0),
        maxi = T)

##
## Linear Programming Results
##
## Call : simplex(a = c(40, 30), A1 = rbind(c(1, 0), c(0, 1), c(1, 2)),
##     b1 = c(16, 8, 24), A2 = rbind(c(1, 0), c(0, 1)), b2 = c(0,
##         0), maxi = T)
##
## Maximization Problem with Objective Function Coefficients
## x1 x2
## 40 30
##
##
## Optimal solution has the following values
## x1 x2
## 16  4
## The optimal value of the objective  function is 760.
```

We also solve the dual of the primal. The dual is given by

$$
\begin{array}{ll}
\text{objective function (min)} & \pi^* = 16y_1 + 8y_2 + 24y_3 + t_1 + t_2 \\
\text{restrictions} & A: 1y_1 + 0y_2 + 1y_3 \geq 40 \\
 & B: 0y_1 + 1y_2 + 2y_3 \geq 30 \\
\text{non-negativity} & y_1, y_2, y_3, t_1, t_2 \geq 0
\end{array}
\tag{5.47}
$$

Because we already use $A2$ and $b2$ for the inequality restrictions of the $\geq$ type, we use $A1$ and $b1$ for the non-negative restrictions by simply changing the sign of the entries of $A1$. The following R code and the results show that the solution values of the maximand of the primal and of the minimand of the dual are identical.

```
library(boot)
simplex(a = c(16,8,24),
        A2 = t(rbind(c(1,0),c(0,1),c(1,2))),
        b2 = c(40,30),
        A1 = rbind(c(-1,0,0),c(0,-1,0),c(0,0,-1)),
        b1 = c(0,0,0))

##
## Linear Programming Results
##
## Call : simplex(a = c(16, 8, 24), A1 = rbind(c(-1, 0, 0), c(0, -1, 0),
##     c(0, 0, -1)), b1 = c(0, 0, 0), A2 = t(rbind(c(1, 0), c(0,
##     1), c(1, 2))), b2 = c(40, 30))
##
## Minimization Problem with Objective Function Coefficients
## x1 x2 x3
## 16  8 24
##
##
```

```
## Optimal solution has the following values
## x1 x2 x3
## 25  0 15
## The optimal value of the objective  function is 760.
```

We work through another simple example. This time we start with a cost minimization problem as the primal. We revisit the cost minimization problem already discussed above.

$$\begin{array}{ll} \text{objective function} & C = 0.6x_1 + x_2 \\ \text{restrictions} & y_1 : 10x_1 + 4x_2 \geq 20 \\ & y_2 : 5x_1 + 5x_2 \geq 20 \\ & y_3 : 2x_1 + 6x_2 \geq 12 \\ \text{non-negativity} & x_1, x_2 \geq 0 \end{array} \tag{5.48}$$

The R-code to set up the problem (primal) and the solution is as follows:

```
library(boot)
simplex(a = c(0.6,1),
        A2 = rbind(c(10,4),c(5,5),c(2,6)),
        b2 = c(20,20,12),
        A1 = rbind(c(-1,0),c(0,-1)),
        b1 = c(0,0))

##
## Linear Programming Results
##
## Call : simplex(a = c(0.6, 1), A1 = rbind(c(-1, 0), c(0, -1)), b1 = c(0,
##     0), A2 = rbind(c(10, 4), c(5, 5), c(2, 6)), b2 = c(20, 20,
##     12))
##
## Minimization Problem with Objective Function Coefficients
##  x1  x2
## 0.6 1.0
##
##
## Optimal solution has the following values
## x1 x2
##  3  1
## The optimal value of the objective  function is 2.8.
```

Again, we solve the dual as well. To obtain the formulation of the dual we use the following correspondence:

$$\begin{array}{lll} \text{Primal} & & \text{Dual} \\ \text{obj. funct. (min) } C = c'x & \longrightarrow & \text{obj. funct. (max) } C^* = r'y \\ \text{restrictions} \quad Ax \geq r & & \text{restrictions} \quad A'y \leq c \\ \text{non-negativity} \quad x \geq 0 & & \text{non-negativity} \quad y \geq 0 \end{array} \tag{5.49}$$

The dual we obtain is:

$$\begin{array}{ll} \text{objective function} & C^* = 20y_1 + 20y_2 + 12y_3 \\ \text{restrictions} & A: 10y_1 + 5y_2 + 2y_3 \leq 0.6 \\ & B: 4y_1 + 5y_2 + 6y_3 \leq 1 \\ \text{non-negativity} & y_1, y_2, y_3 \geq 0 \end{array} \tag{5.50}$$

The solution we find using the following R-code:

```
library(boot)
simplex(a = c(20,20,12),
        A1 = t(rbind(c(10,4),c(5,5),c(2,6))),
        b1 = c(0.6,1),
        A2 = rbind(c(1,0,0),c(0,1,0),c(0,0,1)),
        b2 = c(0,0,0), maxi=T)

##
## Linear Programming Results
##
## Call : simplex(a = c(20, 20, 12), A1 = t(rbind(c(10, 4), c(5, 5), c(2,
##     6))), b1 = c(0.6, 1), A2 = rbind(c(1, 0, 0), c(0, 1, 0),
##     c(0, 0, 1)), b2 = c(0, 0, 0), maxi = T)
##
## Maximization Problem with Objective Function Coefficients
## x1 x2 x3
## 20 20 12
##
##
## Optimal solution has the following values
##   x1   x2   x3
## 0.00 0.08 0.10
## The optimal value of the objective  function is 2.8.
```

5.4 Recommended Readings

This chapter is strongly based on chapters 19 and 20 of

- Chiang AC (1984) Fundamental methods of mathematical economics, 3rd edn. McGraw-Hill, New York.

The exposition in this text is very didactical and works through many examples in great detail.

A seminal reference on optimization is

- Gill PE, Murray W, Wright MH (1991) Numerical linear algebra and optimization, vol. 1 edn. Addison Wesley, London.

5.5 Exercises

1. Consider the following profit maximization problem

$$\begin{array}{ll} \text{objective function} & \pi = 4x_1 + 3x_2 \\ \text{restrictions} & A: 1x_1 + 1x_2 \leq 5 \\ & B: 3x_1 + 2x_2 \leq 12 \\ \text{non-negativity} & x_1, x_2 \geq 0 \end{array}$$

   a. Solve the objective function and the restrictions for x_2 and sketch the problem graphically.
   b. Obtain the point where restrictions A and B intersect.
   c. Obtain the points where the relevant linear restrictions intersect the x- and y-axes.
   d. Indicate the feasible region in your graphic.
   e. Calculate the profits at the three extreme points of the feasible region.
   f. Solve the problem by introducing slack variables, setting up the simplex tableau and solving the problem stepwise by choosing pivot elements and shuffling around the rows.
2. Consider the following cost minimization problem

$$\begin{array}{ll} \text{objective function} & C = 6x_1 + 5x_2 \\ \text{restrictions} & A: 1x_1 + 1x_2 \geq 8 \\ & B: 3x_1 + 2x_2 \geq 18 \\ & C: 1x_1 + 2x_2 \geq 10 \\ \text{non-negativity} & x_1, x_2 \geq 0 \end{array}$$

   a. Solve the objective function and the restrictions for x_2 and sketch the problem graphically.
   b. Obtain the points where the restrictions intersect.
   c. Obtain the points where the relevant linear restrictions intersect the x- and y-axes.
   d. Indicate the feasible region in your graphic.
   e. Calculate the costs at the four extreme points of the feasible region.
   f. Solve the problem by introducing slack variables, setting up the simplex tableau and solving the problem stepwise by choosing pivot elements and shuffling around the rows.
   g. Explain why the origin is no valid starting point for the search of the optimal x_1, x_2-combination and how you circumvent this problem.
3. Consider the profit maximization problem from exercise 1.
   a. Use the function `simplex()` of the boot-package to find the solution.
   b. Derive the dual of the problem.
   c. Solve the dual using the function `simplex()`.

4. Consider the cost minimization problem from exercise 2.
   a. Use the function `simplex()` of the boot-package to find the solution.
   b. Derive the dual of the problem.
   c. Solve the dual using the function `simplex()`.

References

Chiang AC (1984) Fundamental methods of mathematical economics, 3rd edn. McGraw-Hill, New York

Gill PE, Murray W, Wright MH (1991) Numerical linear algebra and optimization, vol. 1 edn. Addison Wesley, London

Data Envelopment Analysis

6

Abstract

Data envelopment analysis is a very popular method to obtain efficiency scores for firms. Its charm is its simplicity. The firms under analysis are compared to the most efficient firm, which most often is a synthetic firm obtained as a linear combination of reference firms. The method is nonparametric as no assumptions on functional relations between inputs and outputs have to be made.

6.1 Introduction

6.1.1 Productivity and Efficiency

The basic idea of productivity is the ratio of output to input

$$\frac{\text{Output}}{\text{Input}}. \tag{6.1}$$

The aim of efficient production is characterized by maximizing output given a certain amount of input or by minimizing input given a certain amount of output. In empirical efficiency analysis, we most often apply a rather relative than absolute concept of efficiency. That is, we do not compare actual output with a theoretical output but with maximum observed output. The empirical analysis is always based on a sample of Decision Making Units (DMUs). The most efficient DMU is regarded as the benchmark for calculating the relative efficiency of the DMUs.[1]

[1] In this chapter we follow rather closely the highly recommended monograph on Data Envelopment Analysis by Cooper et al. (2007).

A. Behr, *Production and Efficiency Analysis with R*,
DOI 10.1007/978-3-319-20502-1_6

6.1.2 One Input–One Output

We start with a very simple example: for each of eight DMUs we have observed the quantity of labor employed and sale. The ratio of output to input is sale per employee. The data is given in Table 6.1 and is generated in R by the following code:

```
emp <- c(2,3,3,4,5,5,6,8)
sales <- c(1,3,2,3,4,2,3,5)
nam <- LETTERS[1:8]
tab <- rbind(emp,sales,round(sales/emp,3))
colnames(tab) <- nam
rownames(tab) <- c("Employees","Sales","Sales/Empl.")
```

To the different DMUs we refer by capital letters. We find that *B* has the highest ratio of output to input. Therefore, DMU *B* is the unit of reference. If we assume constant returns to scale and that DMUs can simply be scaled up or down, a straight line from the origin through *B* is the "efficient frontier". For example a DMU half the size of *B* will be able to produce half the output of *B* and also a DMU twice the size of *B* will be able to produce twice the output of *B*.

We can benchmark all DMUs indexed by *i* relative to the efficient DMU *B*

$$0 \leq \frac{\text{Sales per employee of DMU } i}{\text{Sales per employee of B}} \leq 1. \tag{6.2}$$

The relative efficiency describes an aspect of production for a specific period of time. An inefficient DMU can become efficient by moving to the frontier. This can be achieved either through increasing output at constant input or through decreasing input at constant output. Of course, inputs and outputs can also change simultaneously.

We can display the situation with the following code (see Fig. 6.1).

```
# plot data
plot(emp,sales,pch=19,cex=0.8,
     xlim=c(0,9), ylim=c(0,6),
     xlab="Employees", ylab="Sales")
# labels
text(emp+0.4,sales+0.1,nam)
# efficient frontier
segments(0,0,6,6)
```

Table 6.1 One input–one output production data

	A	B	C	D	E	F	G	H
Employees	2.00	3.00	3.00	4.00	5.00	5.00	6.00	8.00
Sales	1.00	3.00	2.00	3.00	4.00	2.00	3.00	5.00
Sales/Empl.	0.50	1.00	0.67	0.75	0.80	0.40	0.50	0.62

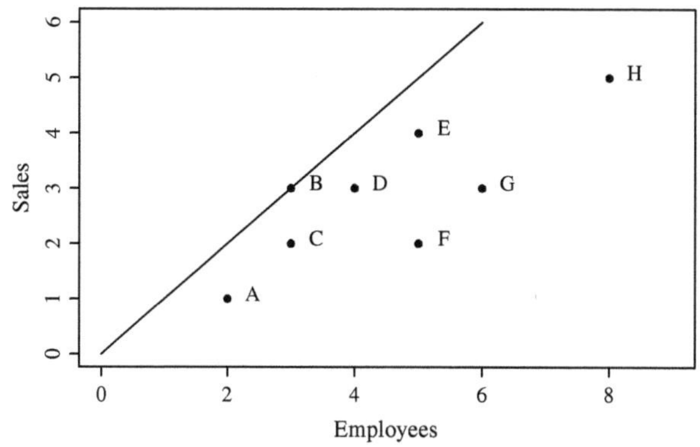

Fig. 6.1 One input–one output case

Table 6.2 Two inputs–one output production data

	A	B	C	D	E	F	G
Employee	4.00	7.00	8.00	4.00	2.00	5.00	6.00
Floor Area	3.00	3.00	1.00	2.00	4.00	2.00	4.00
Sales	1.00	1.00	1.00	1.00	1.00	1.00	1.00

6.1.3 Two Inputs–One Output

We now turn to a situation with two inputs (employees and floor area) and one output (sales). We express both inputs relative to output, obtaining average input per unit of output. Using this ratio implies that we are again assuming constant returns to scale. Using a 2-dimensional diagram, the piecewise linear production possibility set can be displayed. The "efficient frontier" assumes that two DMUs can be linearly combined (convexity). The data for the simple numerical example are given in Table 6.2 and is generated in R by the following code:

```
nam <- LETTERS[1:7]
emp <- c(4,7,8,4,2,5,6)
flo <- c(3,3,1,2,4,2,4)
sales <- rep(1,7)
tab <- round(rbind(emp,flo,sales),3)
colnames(tab) <- nam
rownames(tab) <- c("Employee","Floor Area","Sales")
```

We can display the situation with the following code (see Fig. 6.2).

```
# plot data
plot(emp,flo,pch=19,cex=0.8,
     xlim=c(0,9), ylim=c(0,5),
     xlab="Employee/Sales", ylab="Area/Sales")
# efficient frontier
lines(emp[5:3],flo[5:3])
segments(emp[3],flo[3],emp[3]+1,flo[3])
```

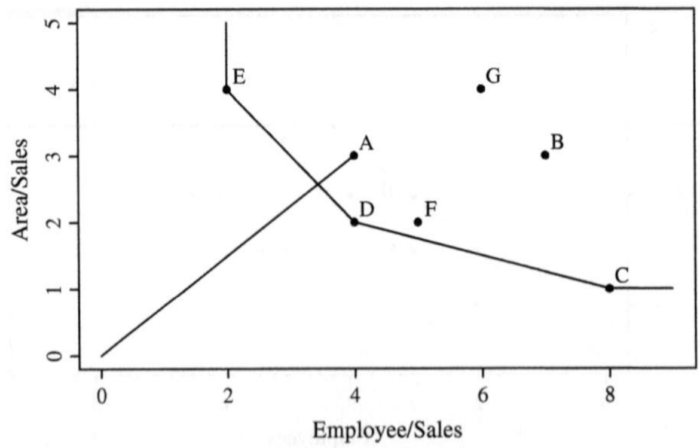

Fig. 6.2 Two inputs–one output case

Table 6.3 One input–two outputs production data

	A	B	C	D	E	F	G
Employees	1.00	1.00	1.00	1.00	1.00	1.00	1.00
Customers	1.00	2.00	3.00	4.00	4.00	5.00	6.00
Sales	5.00	7.00	4.00	3.00	6.00	5.00	2.00

```
segments(emp[5],flo[5],emp[5],flo[5]+1)
# labels
text(emp+0.2,flo+0.2,nam)
# inefficient unit A
segments(0,0,emp[1],flo[1])
```

The three efficient units C, D, and E serve as the benchmark for the non-efficient DMUs. If we, for example, look at inefficient unit A, we find that the two relevant efficient units D and E form the "reference set" for unit A. Note that all linear combinations of the reference set $\{D, E\}$ are efficient. If unit A moves to the efficient frontier, it will become efficient.

There are several ways for DMU A to reach the efficient frontier: A could reduce either the input area or the input employees, A could reduce both inputs when moving along the path through the origin towards the origin, A could maintain the amount of inputs and increase output.

6.1.4 One Input–Two Outputs

The data for the small example with one input (employees) and two outputs (customers and sales) are given in Table 6.3 and is generated in R by the following code:

```
cus <- c(1:4,4:6)
sales <- c(5,7,4,3,6,5,2)
emp <- rep(1,7)
```

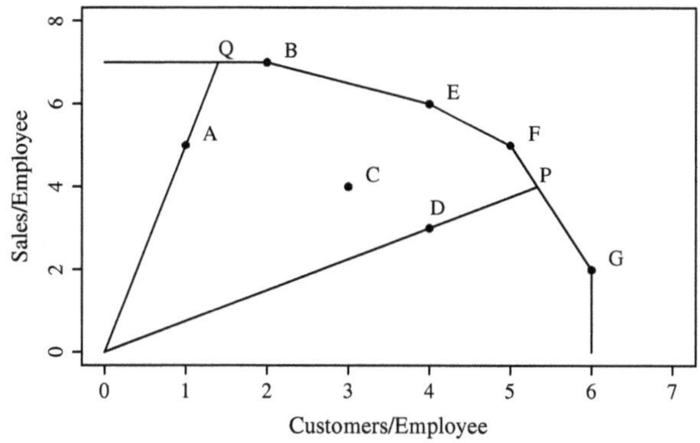

Fig. 6.3 One input–two outputs case

```
nam <- LETTERS[1:7]
tab <- round(rbind(emp,cus,sales),3)
colnames(tab) <- nam
rownames(tab) <- c("Employees","Customers","Sales")
```

The outputs are standardized per unit of input. Using average amount of output per input again implies constant returns to scale. In Fig. 6.3, we observe that the DMUs B, E, F, and G are efficient and the linear connections form the "efficient frontier". To create the plot, we use the following code:

```
# plot data
plot(cus,sales,pch=19,cex=0.8,
     xlim=c(0,7), ylim=c(0,8),
     xlab="Customers/Employee", ylab="Sales/Employee")
# labels
text(cus[-4]+0.3,sales[-4]+0.3,nam[-4])
text(cus[4]+0.1,sales[4]+0.5,nam[4])
# efficient frontier
lines(cus[-c(1,3,4)],sales[-c(1,3,4)])
segments(cus[2],sales[2],0,sales[2])
segments(cus[7],sales[7],cus[7],0)
# inefficient units A and D
segments(0,0,1.4,7)
text(1.4+0.1,7+0.3,"Q")
segments(0,0,16/3,4)
text(16/3+0.1,4+0.3,"P")
```

Looking at DMU D, we observe that the efficient units F and G form the reference for D. Again, we assume that all linear combinations of the two efficient DMUs are a relevant benchmark. The line from the origin through D meets the frontier in point P. Note that P is a synthetic DMU and its relevance rests on the assumption of convexity. Moving the inefficient unit D towards the efficient point of production P

implies an increase in both outputs using the same amount of input. Comparing the length d(0, D) with d(0, P) serves as a measure of inefficiency

$$\frac{\mathrm{d}(0, D)}{\mathrm{d}(0, P)} = 0.75. \tag{6.3}$$

Hence, actually unit D produces only 75 % of the possible output under efficient production. Analogously, the output could be increased by the factor of $4/3 = 1.333$. The coordinates of the efficient point P can be obtained from the coordinates of D

$$\frac{4}{3}(4, 3) = \left(\frac{16}{3}, 4\right). \tag{6.4}$$

6.1.5 Fixed and Variable Weights

So far, we only discussed examples with either one input or one output. In these cases, we could standardize the outputs or inputs using the single input or single output. This method is not applicable in the case of several inputs and outputs. A simple comparison of output relative to input is only possible if we can sum up the different inputs towards a single measure of input and sum up the different outputs towards a single measure of output. As the different goods cannot be summed up using their quantities, we have to use weights (prices).

We consider the simple 2 input–2 output case. Example data are given in Table 6.4.

The two inputs are doctors (x_1) and nurses (x_2) and the two outputs are outpatients (y_1) and inpatients (y_2). More generally, we denote outputs with y_r $(r = 1, \ldots, s)$, inputs with x_i $(i = 1, \ldots, m)$, output prices with u_r $(r = 1, \ldots, s)$, and input prices with v_i $(i = 1, \ldots, m)$. j is the index for the DMUs $(j = 1, \ldots, n)$. Given input prices v_1 and v_2 and output prices u_1 and u_2, we can compare the sum of weighted output with the sum of weighted input

$$\frac{\text{Output}}{\text{Input}} = \frac{\sum_r y_r u_r}{\sum_i x_i v_i}. \tag{6.5}$$

Note that scaling output prices $c \cdot u_r$ and input prices $k \cdot v_i$ does not alter the efficiency measure if comparing different DMUs. Hence, only relative input and output prices

Table 6.4 Two inputs–two outputs production data

	A	B	C	D	E	F	G	H
x1: Doctors	20	19	25	27	22	55	33	31
x2: Nurses	151	131	160	168	158	255	235	206
y1: Outpatients	100	150	160	180	94	230	220	152
y2: Inpatients	90	50	55	72	66	90	88	80

Table 6.5 Two inputs–two outputs production data, effect of different weighting schemes

	A	B	C	D	E	F	G	H
x1: Doctors	20.00	19.00	25.00	27.00	22.00	55.00	33.00	31.00
x2: Nurses	151.00	131.00	160.00	168.00	158.00	255.00	235.00	206.00
y1: Outpat.	100.00	150.00	160.00	180.00	94.00	230.00	220.00	152.00
y2: Inpat.	90.00	50.00	55.00	72.00	66.00	90.00	88.00	80.00
Weights 1	1.47	1.33	1.14	1.31	1.09	0.94	1.21	1.09
Weights 2	0.99	1.18	1.02	1.14	0.79	0.88	1.02	0.87
Efficiency 1	1.00	0.90	0.77	0.89	0.74	0.64	0.82	0.74
Efficiency 2	0.84	1.00	0.87	0.96	0.67	0.74	0.86	0.73

are relevant. Of course, the resulting measure of efficiency depends on the chosen relative input and output prices.

Table 6.5 contains two different (arbitrarily chosen) weights. Weighting scheme 1:

$$\frac{v_1}{v_2} = \frac{5}{1}; \; \frac{u_1}{u_2} = \frac{1}{3}. \tag{6.6}$$

Weighting scheme 2:

$$\frac{v_1}{v_2} = \frac{2}{1}; \; \frac{u_1}{u_2} = \frac{1}{1}. \tag{6.7}$$

We observe the efficiency scores to depend strongly on the weighting scheme used. For measuring efficiency, we therefore have to decide which weighting scheme to use.

6.2 Charnes–Cooper–Rhodes-Model

6.2.1 Introduction

The ratio of weighted outputs to weighted inputs depends on the weighting scheme chosen. So how to judge the efficiency of DMUs if their relative efficiency depends on the weighting scheme? The idea of the Charnes–Cooper–Rhodes-model (CCR-Model) is to choose for each DMU the set of input and output prices which results in the maximum ratio of weighted output to weighted input given a set of restrictions:

1. all input and output prices must be non-negative.
2. for all DMUs weighted output must not exceed weighted input.

For the DMU under consideration o, we choose v and u to maximize θ under the constraint that for all n DMUs weighted output does not exceed weighted input and all prices are non-negative. Therefore, the problem is expressed as a linear programming problem which can be solved by means of the simplex algorithm. Obviously, the number of restrictions which have to be met is rather large. The number equals the number of firms (n) plus the number of inputs (m) plus the number of outputs (s).

This maximization problem has to be solved for each of the n DMUs and in each maximization problem $n + m + s$ constraints have to be met. For firm o the problem is formally given as:

$$
\begin{aligned}
(LP_o) \qquad & \max_{v,u} \theta = \frac{\sum_r y_{ro} u_r}{\sum_i x_{io} v_i} \\
\text{subject to } & \frac{\sum_r y_{rj} u_r}{\sum_i x_{ij} v_i} \leq 1 \ (j = 1, \ldots, n) \\
& v_i \geq 0 \ (i = 1, \ldots, m) \\
& u_r \geq 0 \ (r = 1, \ldots, s).
\end{aligned}
\tag{6.8}
$$

6.2.2 Programming Problem

We consider positive input prices $v > 0$ and positive amounts of input $x > 0$ and normalize the input from DMU o to 1. The maximization problem then is

$$
\begin{aligned}
(LP_o) \qquad & \max_{v,\mu} \theta = \sum_r y_{ro} \mu_r \\
\text{subject to } & \sum_i x_{io} v_i = 1 \\
& \sum_r y_{rj} \mu_r \leq \sum_i x_{ij} v_i \ (j = 1, \ldots, n) \\
& v_i \geq 0 \ (i = 1, \ldots, m) \\
& u_r \geq 0 \ (r = 1, \ldots, s).
\end{aligned}
\tag{6.9}
$$

θ^* denotes the solution to the maximization problem. v^* and u^* are the vectors of optimal input and output prices. A DMU is efficient only if $\theta^* = 1$. In this case, its weighted output equals its weighted input and the restriction that weighted output must be less equal weighted input is just met. Beside that restriction, we have the non-negativity constraints $v^* \geq 0$ and $u^* \geq 0$. If $\theta^* < 1$, then at least for one DMU (usually for several) we find $\sum_r y_r u_r^* = \sum_i x_i v_i^*$. DMUs for which this equality holds belong to the reference set E'_o of the inefficient DMU o

$$
E'_o = \left\{ j : \sum_r y_r u_r^* = \sum_i x_i v_i^* \right\}. \tag{6.10}
$$

The set of efficient DMUs in E'_o span the "efficient frontier" for inefficient DMU o. Hence, all inefficient DMUs are measured relative to their specific reference sets.

Table 6.6 One input–one output case

	A	B	C	D	E	F	G	H
x: Input	2.000	3.000	3.000	4.000	5.000	5.000	6.000	8.000
y: Output	1.000	3.000	2.000	3.000	4.000	2.000	3.000	5.000
A: u ≤	1.000	0.500	0.750	0.667	0.625	1.250	1.000	0.800
B: u ≤	0.667	0.333	0.500	0.444	0.417	0.833	0.667	0.533

6.2.3 Examples

In this section, we work through some simple examples. We start with the most simple case considering only one input and one output. The data is given in Table 6.6.

6.2.3.1 One Input–One Output

When standardizing the weighted input towards 1, we obtain the input price v as the reciprocal of the input quantity. For firm A, e.g., with input quantity 2 we obtain $v^* = 0.5$. As the output quantity of firm A is just 1, maximizing weighted output is simply maximizing the output price u. Meeting the restriction that for no firm weighted output must exceed weighted input, maximal u for firm A is obtained as the minimum u which guarantees that the restrictions are met for all firms. In this example, the restriction for firm B $3u \leq 3v$ results in $u \leq 0.5$ for $v^* = 0.5$. Formally, the programming problem for DMU A is given as

$$\begin{aligned} &< A > \ \max \theta = u \\ &\text{subject to} \quad 2v = 1 \\ &\qquad (A)\ u \leq 2v \quad (B)\ 3u \leq 3v \\ &\qquad (C)\ 2u \leq 3v \quad (D)\ 3u \leq 4v \\ &\qquad (E)\ 4u \leq 5v \quad (F)\ 2u \leq 5v \\ &\qquad (G)\ 3u \leq 6v \quad (H)\ 5u \leq 8v. \end{aligned} \tag{6.11}$$

The smallest u is the solution, which because of output quantity 1, is also the solution for θ: $v^* = 0.5$, $u = 0.5$, $\theta^* = 0.5$. Therefore, if we multiply the inputs of the inefficient DMU A with factor $\theta^* = 0.5$, we obtain the efficient amount of input for the given output.

The LP-problem for DMU B is given as

$$\begin{aligned} &< B > \ \max \theta = 3u \\ &\text{subject to} \quad 3v = 1 \\ &\qquad (A)\ u \leq 2v \quad (B)\ 3u \leq 3v \\ &\qquad (C)\ 2u \leq 3v \quad (D)\ 3u \leq 4v \\ &\qquad (E)\ 4u \leq 5v \quad (F)\ 2u \leq 5v \\ &\qquad (G)\ 3u \leq 6v \quad (H)\ 5u \leq 8v. \end{aligned} \tag{6.12}$$

Table 6.7 Two inputs–one output

	A	B	C	D	E	F	G
x1: Input	4.00	7.00	8.00	4.00	2.00	10.00	3.00
x2: Input	3.00	3.00	1.00	2.00	4.00	1.00	7.00
y: Output	1.00	1.00	1.00	1.00	1.00	1.00	1.00

We first find that $v^* = 1/3$. Given this value, we can solve all restrictions for u. Because all restrictions have to be met, the smallest u is the solution: $v^* = 1/3$, $u = 1/3$, $\theta^* = 1$. Solution $\theta^* = 1$ means that firm B is efficient. Therefore, if we multiply the inputs of the efficient DMU B with factor $\theta^* = 1$, we obtain the efficient amount of input for the given output, which is just the amount actually used by firm B. Table 6.6 contains two rows showing the values of u resulting from the eight restrictions. The smallest u is the maximal value of u that can be obtained in the programming problem for firm A (the last but one row) and firm B (last row).

6.2.3.2 Two Inputs–One Output

We consider a simple example with two inputs and one output (see Table 6.7). The LP-problem for DMU A is given as

$$\begin{aligned} &< A > \max \theta = u \\ &\text{subject to} \quad 4v_1 + 3v_2 = 1 \\ &\qquad (A)\ u \le 4v_1 + 3v_2 \quad (B)\ u \le 7v_1 + 3v_2 \\ &\qquad (C)\ u \le 8v_1 + v_2 \quad (D)\ u \le 4v_1 + 2v_2 \\ &\qquad (E)\ u \le 2v_1 + 4v_2 \quad (F)\ u \le 10v_1 + v_2 \\ &\qquad (G)\ u \le 3v_1 + 7v_2. \end{aligned} \tag{6.13}$$

The LP-problem can be solved using the simplex-algorithm. We have to find u, v_1 and v_2 which maximize the target function $1u + 0v_1 + 0v_2$ meeting the restrictions. The solution for A is: $u^* = 0.8751, v_1^* = 0.1429, v_2^* = 0.1429, \theta^* = 0.8751$. We find that firm A has an efficiency score of 0.8751. Therefore, this unit could reduce its input vector by 12.5 %. The corresponding LP-problems have to be solved for all n DMUs to obtain efficiency scores for the n DMUs.

6.2.4 Production Correspondence

We assume non-negative input vector $\mathbf{x}_j$ and non-negative output vector $\mathbf{y}_j$. The pair of input and output vector $(\mathbf{x}_j, \mathbf{y}_j)$ with $\mathbf{x}_j \in R^m$ and $\mathbf{y}_j \in R^s$ is called "activity". The set of potential activities $\mathscr{P}$ is called "production possibility set". The set $\mathscr{P}$ has the following characteristics:

1. Observed activities $(\mathbf{x}_j, \mathbf{y}_j)$ $(j = 1, \ldots, n)$ belong to $\mathscr{P}$.
2. If $(\mathbf{x}, \mathbf{y})$ belongs to $\mathscr{P}$, then $(t\mathbf{x}, t\mathbf{y})$ belongs to $\mathscr{P}$ for a positive scalar t $(t > 0)$. This property is the constant-returns-to-scale assumption.

3. If $(\mathbf{x}, \mathbf{y})$ belongs to $\mathscr{P}$, then any semi-positive activity $(\bar{\mathbf{x}}, \bar{\mathbf{y}})$ with $\bar{\mathbf{x}} \geq \mathbf{x}$ and $\bar{\mathbf{y}} \leq \mathbf{y}$ belongs to $\mathscr{P}$.
4. Any semi-positive linear combination of observed activities belongs to $\mathscr{P}$ (convexity).

If we observe n firms with activities $(\mathbf{x}_j, \mathbf{y}_j)$ $(j = 1, \ldots, n)$, we can combine the inputs $\mathbf{x}_j$ column wise to obtain a matrix X with dimension $(m \times n)$ and the outputs $\mathbf{y}_j$ column wise to obtain a matrix Y with dimension $(s \times n)$. Using $\boldsymbol{\lambda}$, which is a semi-positive column vector of dimension $(n \times 1)$, we can define the production possibility set $\mathscr{P}$ as follows:

$$\begin{aligned} P &= \{(\mathbf{x}, \mathbf{y}) | \mathbf{x} \geq X\boldsymbol{\lambda}, \mathbf{y} \leq Y\boldsymbol{\lambda}, \boldsymbol{\lambda} \geq 0\} \\ X &= (\mathbf{x}_1, \ldots, \mathbf{x}_n); Y = (\mathbf{y}_1, \ldots, \mathbf{y}_n); \boldsymbol{\lambda} \in R^n. \end{aligned} \tag{6.14}$$

Note that $\boldsymbol{\lambda} = (\lambda_1, \ldots, \lambda_j, \ldots, \lambda_n)'$ contains the weights, which are given to the inputs and outputs of the n firms when constructing a synthetical firm as reference with input $X\boldsymbol{\lambda}$ and output $Y\boldsymbol{\lambda}$. $X\boldsymbol{\lambda}$ is the column vector containing the m inputs of this synthetical reference firm. For each component x_i of the input vector of the activity $\mathbf{x}, \mathbf{y}$ it must hold that it is not smaller than the corresponding quantity (i-th row) of $X\boldsymbol{\lambda}$ (i.e., sum of weighted inputs x_{ij} which is $\sum_j \lambda_j x_{ij}$). Similarly, each component y_r of the output vector of the activity $\mathbf{x}, \mathbf{y}$ must not exceed the r-th row of $Y\boldsymbol{\lambda}$ (i.e., sum of weighted outputs y_{rj} which is $\sum_j \lambda_j y_{rj}$). Therefore, $\mathscr{P}$ must not contain activities which are more efficient than linear combinations of observed activities.

So far, we considered the maximization problem using input-multipliers v (fictive input-prices) and output-multipliers u (fictive output-prices). Hence, the resulting linear programming problem is said to have the multiplier form:

$$\begin{aligned} &(LP_o) \quad \max_{\mathbf{v},\mathbf{u}} \mathbf{u}'\mathbf{y}_o \\ &\text{subject to} \quad \mathbf{v}'\mathbf{x}_o = 1 \\ &\qquad -\mathbf{v}'X + \mathbf{u}'Y \leq 0 \\ &\qquad \mathbf{v} \geq 0, \ \mathbf{u} \geq 0. \end{aligned} \tag{6.15}$$

The dual of the linear program (LP) in multiplier form is called the LP in envelopment form:

$$\begin{aligned} &(DLP_o) \quad \min_{\theta,\boldsymbol{\lambda}} \theta \\ &\text{subject to} \quad \theta\mathbf{x}_o - X\boldsymbol{\lambda} \geq 0 \\ &\qquad Y\boldsymbol{\lambda} \geq \mathbf{y}_o \\ &\qquad \boldsymbol{\lambda} \geq 0. \end{aligned} \tag{6.16}$$

Whereas in the multiplier form we choose the vectors of input and output prices to maximize sales, we now choose the vector $\boldsymbol{\lambda}$ which combines the observed activities of the n firms and θ which states the factor of input reduction possible when adopting the efficient production technique. $X\boldsymbol{\lambda}$ denotes the input vector of the combined activity and $Y\boldsymbol{\lambda}$ denotes the output vector of the combined activity. Under efficient production, the observed input quantity $\mathbf{x}_o$ can be reduced towards $\theta\mathbf{x}_o$. In the combined production process the amount of input $X\boldsymbol{\lambda}$ must not exceed the amount used in the optimum $\theta\mathbf{x}_o$ and the amount of output $Y\boldsymbol{\lambda}$ must not fall below observed output $\mathbf{y}_o$. The solution θ^* $(0 \leq \theta^* \leq 1)$ indicates the fraction of the observed amount of input $\mathbf{x}_o$ needed if producing the output $\mathbf{y}_o$ efficiently. The "proportional" reduced amount of input $\theta\mathbf{x}_o$ for producing observed output y_o can imply wasted input s^-

$$\mathbf{s}^- = \theta\mathbf{x}_o - X\boldsymbol{\lambda} \tag{6.17}$$

or forgone output $\mathbf{s}^+$

$$\mathbf{s}^+ = Y\boldsymbol{\lambda} - \mathbf{y}_o \tag{6.18}$$

where $\mathbf{s}^- \geq 0$ and $\mathbf{s}^+ \geq 0$.

The observed input mix used by an inefficient DMU can be proportionally reduced by the factor $(1 - \theta^*)$ when moving towards the efficient frontier. Nevertheless, there might be some slacks. Slacks in the production process are present if at least one input could be reduced further without decreasing output or if at least one output could be increased given the input. This problem is analyzed in a two-step approach:

1. The LP-problem is solved to obtain the optimal θ denoted by θ^*.
2. Given $\theta = \theta^*$, we search for the optimal value of $\boldsymbol{\lambda}$, which maximizes the input slacks $\mathbf{s}^- = \theta\mathbf{x}_o - X\boldsymbol{\lambda}$ or foregone output $\mathbf{s}^+ = Y\boldsymbol{\lambda} - \mathbf{y}_o$.

The second step again is an LP-problem

$$\begin{aligned} \max_{\boldsymbol{\lambda},\mathbf{s}^+,\mathbf{s}^-} \quad & \omega = \mathbf{e}\mathbf{s}^+ + \mathbf{e}\mathbf{s}^- \\ \text{subject to } & \mathbf{s}^- = \theta\mathbf{x}_o - X\boldsymbol{\lambda} \\ & \mathbf{s}^+ = Y\boldsymbol{\lambda} - \mathbf{y}_o \\ & \boldsymbol{\lambda} \geq 0, \mathbf{s}^- \geq 0, \mathbf{s}^+ \geq 0. \end{aligned} \tag{6.19}$$

To characterize the DMU as CCR-efficient, the following equalities must hold:

$$\theta^* = 1, \quad \mathbf{s}^- = 0, \quad \mathbf{s}^+ = 0. \tag{6.20}$$

Therefore, a DMU is said to be CCR-efficient if there is no radial reduction in input possible and there are no slacks. The potential existence of slacks even after a radial reduction in input is the distinction between two different concepts of efficiency:

- If $\theta^* = 1$ but $\mathbf{s}^- \gneq 0$ or $\mathbf{s}^+ \gneq 0$, the DMU is Farrell-efficient.[2]
- If $\theta^* = 1$ and $\mathbf{s}^- = 0, \mathbf{s}^+ = 0$, the DMU is Pareto–Koopman-efficient.

The computational scheme of the CCR-Model can be written as a two-stage LP problem

$$\begin{array}{lll} & (DLP_o) & \\ \text{Phase I objective} & \min & \theta \\ \text{Phase II objective} & \min & -\mathbf{e}\mathbf{s}^+ - \mathbf{e}\mathbf{s}^- \\ \text{subject to} & & \theta \mathbf{x}_o = X\boldsymbol{\lambda} + \mathbf{s}^- \\ & & \mathbf{y}_o = Y\boldsymbol{\lambda} - \mathbf{s}^+ \\ & & \theta \geq 0, \boldsymbol{\lambda} \geq 0, \mathbf{s}^- \geq 0, \mathbf{s}^+ \geq 0. \end{array} \tag{6.21}$$

In more general LP notation the problem can be stated as follows

$$\begin{array}{lll} & (DLP'_o) & \\ \text{Phase I objective} & \min & z_1 = \mathbf{c}\mathbf{x} \\ & \min & z_2 = \mathbf{d}\mathbf{x} \\ \text{subject to} & & Ax = \mathbf{b} \\ & & x \geq \mathbf{0} \end{array} \tag{6.22}$$

where

$$\begin{aligned} &\mathbf{x} = (\theta, \boldsymbol{\lambda}', \mathbf{s}^{-\prime}, \mathbf{s}^{+\prime})', \quad \mathbf{c} = (1, \mathbf{0}, \mathbf{0}, \mathbf{0}), \quad \mathbf{d} = (0, \mathbf{0}, -\mathbf{e}, -\mathbf{e}), \\ &A = \left(\begin{bmatrix} \mathbf{x}_o & -X & -I & 0 \\ \mathbf{0} & Y & 0 & -I \end{bmatrix} \right) \quad \text{and} \quad \mathbf{b} = \left(\begin{bmatrix} \mathbf{0} \\ \mathbf{y}_o \end{bmatrix} \right). \end{aligned} \tag{6.23}$$

We return to our last numerical example (Table 6.7). The two input–one output example can be displayed graphically (Fig. 6.4). We find that firms E, D, C, F are Farrell-efficient. For these firms, no radial reduction in input is possible if their output level should be maintained. Firms G, A, B are inefficient and therefore can proportionally reduce their amount of inputs holding output constant. If firms A and B would radially reduce their inputs, they would become Farrell- as well as Koopman-efficient. Firm F is Farrell-efficient, but not Koopman-efficient, because there is some slack in input x_1. Input x_1 could be reduced by the amount of 2 holding output constant. If firm G radially reduces its input, it will shift onto the efficient frontier and therefore then will be Farrell-efficient. As there would still be some slack in input x_2, it still would not be Koopman-efficient. Note that inefficient firms are compared to synthetic firms. For example firm A is compared with a synthetic

[2]The symbol $\gneq$ indicates that for all elements of the vector $\geq$ and at least for one element $>$ holds.

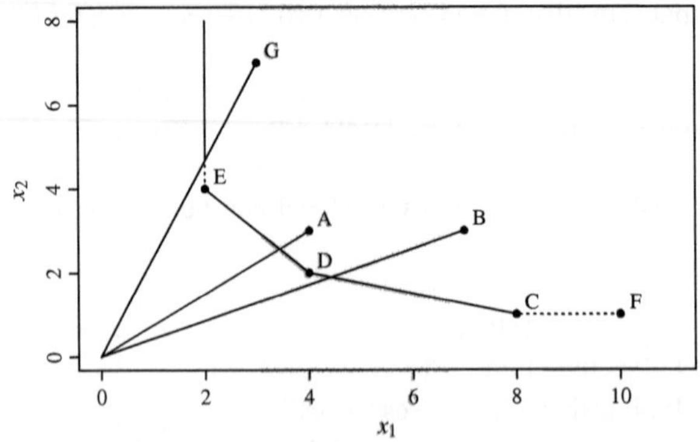

Fig. 6.4 Two input–one output case

Table 6.8 Two inputs–one output case: results of DEA analysis assuming constant returns to scale

	A	B	C	D	E	F	G
x1: Input	4.00	7.00	8.00	4.00	2.00	10.00	3.00
x2: Input	3.00	3.00	1.00	2.00	4.00	1.00	7.00
y: Output	1.00	1.00	1.00	1.00	1.00	1.00	1.00
Efficiency	0.86	0.63	1.00	1.00	1.00	1.00	0.67
Slack x1	0.00	0.00	0.00	0.00	0.00	2.00	0.00
Slack x2	0.00	0.00	0.00	0.00	0.00	0.00	0.67
$A : \lambda$	0.00	0.00	0.00	0.71	0.29	0.00	0.00
$B : \lambda$	0.00	0.00	0.11	0.89	0.00	0.00	0.00
$G : \lambda$	0.00	0.00	0.00	0.00	1.00	0.00	0.00

firm consisting of 71 % of firm D and 29 % of firm E. All the results are given in Table 6.8.

6.2.5 Output-Oriented Model

So far, we analyzed the input-oriented model, that is we minimized input while producing at least the observed output level. In the output-oriented model the aim is to maximize output while using no more than the observed inputs:

$$
\begin{aligned}
(DLPO_o) \quad & \max_{\eta,\mu} \ \eta \\
\text{subject to} \quad & \mathbf{x}_o - X\boldsymbol{\mu} \geq 0 \\
& \eta \mathbf{y}_o - Y\boldsymbol{\mu} \leq 0 \\
& \boldsymbol{\mu} \geq 0.
\end{aligned}
\tag{6.24}
$$

Note that there is a direct correspondence between the input- and the output-oriented model. Having obtained the solution of the input-oriented model, the solution of the output-oriented model can easily be derived. The following relation holds

$$\boldsymbol{\lambda} = \frac{\boldsymbol{\mu}}{\eta}, \quad \theta = \frac{1}{\eta}. \tag{6.25}$$

Inserting this relation into the output-oriented model results in the input-oriented model

$$\begin{aligned} (DLP_o) \quad & \min_{\theta,\boldsymbol{\lambda}} \theta \\ \text{subject to} \quad & \theta \mathbf{x}_o - X\boldsymbol{\lambda} \geq 0 \\ & \mathbf{y}_o - Y\boldsymbol{\lambda} \leq 0 \\ & \boldsymbol{\lambda} \geq 0. \end{aligned} \tag{6.26}$$

Therefore, the following relation holds also for the optimal values

$$\boldsymbol{\lambda}^* = \frac{\boldsymbol{\mu}^*}{\eta^*}, \quad \theta^* = \frac{1}{\eta^*}. \tag{6.27}$$

Instead of noting the slacks $\mathbf{s}^- = \theta\mathbf{x}_o - X\boldsymbol{\lambda}$ and $\mathbf{s}^+ = Y\boldsymbol{\lambda} - \mathbf{y}_o$, we now denote the slacks as

$$\mathbf{t}^- = \mathbf{x}_o - X\boldsymbol{\mu}, \quad \mathbf{t}^+ = Y\boldsymbol{\mu} - \eta\mathbf{y}_o. \tag{6.28}$$

For the slacks there is a direct link between input- and output-oriented models

$$\mathbf{t}^{-*} = \frac{\mathbf{s}^{-*}}{\theta^*}, \quad \mathbf{t}^{+*} = \frac{\mathbf{s}^{+*}}{\theta^*}. \tag{6.29}$$

6.2.6 Returns to Scale in DEA Models

We restrict the discussion of different assumptions on returns to scale to a simple one input–one output example. Consider five data points (input,output): $A = \{2, 1\}, B = \{3, 3\}, C = \{5, 4\}, D = \{2.5, 1.5\}, E = \{4, 2.5\}$.

The first stage problem of the CCR-model with input minimization had been stated as follows:

$$\begin{aligned} & (DLP_o) \\ \text{objective min} \quad & \theta \\ \text{subject to} \quad & \theta\mathbf{x}_o = X\boldsymbol{\lambda} + \mathbf{s}^- \\ & \mathbf{y}_o = Y\boldsymbol{\lambda} - \mathbf{s}^+ \\ & \theta \geq 0, \boldsymbol{\lambda} \geq 0, \mathbf{s}^- \geq 0, \mathbf{s}^+ \geq 0. \end{aligned} \tag{6.30}$$

We only apply the restriction $\boldsymbol{\lambda} \geq 0$, which allows to scale inputs and outputs downward $\sum \lambda < 1$ or upward $\sum \lambda > 1$. Note that in this case B is the reference set, hence $\sum \lambda$ consists of only one summand which is λ_B. We illustrate this situation graphically (see Fig. 6.5a).

For example the hypothetical input–output combination that firm A is compared to is A' which is B scaled down with $\lambda = 1/3$ and C is compared to C' which is B scaled up by $\lambda = 4/3$.

Under the assumption of decreasing returns to scale we allow firms only to be scaled down (see Fig. 6.5b). Therefore, point C' does not belong to the frontier which now will run from 0 to B and from B to C, hence C now is regarded efficient. In this case, E will be compared to a hypothetical input–output combination E', which is obtained as a linear combination with equal weight (0.5) of B and C. The linear program in the case of decreasing returns to scale is supplemented with the condition $\sum \lambda \leq 1$.

Under the assumption of increasing returns to scale we allow firms only to be scaled up (see Fig. 6.6a). Therefore, the segment running from the origin to B does not belong to the frontier which now will run from point $(2, 0)$ to A, from A to B and from there on it is the path from the origin through B. Hence, D will now be compared to a hypothetical input–output combination D', which is obtained as a

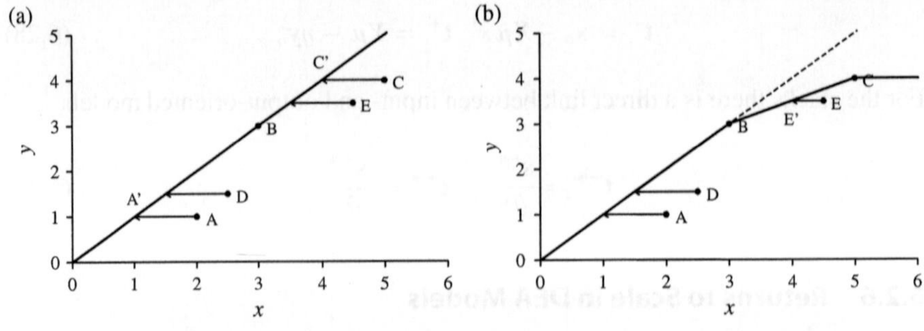

Fig. 6.5 Returns to scale (1). **(a)** Constant returns to scale. **(b)** Decreasing returns to scale

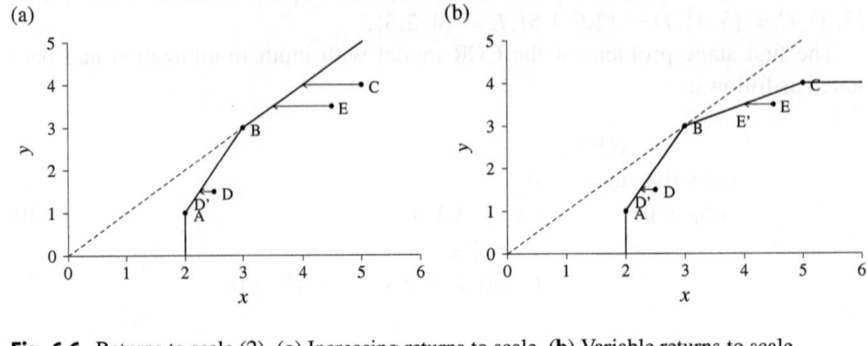

Fig. 6.6 Returns to scale (2). **(a)** Increasing returns to scale. **(b)** Variable returns to scale

linear combination of A (weight 0.75) and B (weight 0.25). The linear program in the case of increasing returns to scale is supplemented with the condition $\sum \lambda \geq 1$. In this case D and E both are compared to hypothetical firms.

The most restrictive case we consider is the case of variable returns to scale (see Fig. 6.6b). In this case firms are always compared to a hypothetical firm that is a linear combination of firms whereof at least one is smaller and at least one is bigger in terms of output (the exception is the rare case of firms with identical output, in which case the reference set consists of only one firm). This prevents rather unrealistic comparisons of firms of strongly different size which can become relevant in the other cases. Think about the relevance of comparing a big firm with thousands of employees with a small "efficient" firm that had to be scaled up by a factor of 100. In the case of variable returns to scale the weights of the firms have to meet the restriction $\sum \lambda = 1$. In this case D and E both are compared to hypothetical firms which are linear combinations, i.e., of A and B and of B and C respectively.

6.3 Data Envelopment Analysis with R

6.3.1 Benchmarking Package

In this section, we use the Benchmarking package, which contains a rather extensive number of methods for parametric and nonparametric efficiency analysis.[3] The package offers a variety of DEA methods as well as easy to use plotting facilities. In this introductory chapter, we only cover the most basic functions of this extensive package.

6.3.2 Examples

We work through three different examples, the one input–one output case, the two inputs–one output case, and the one input–two outputs case. While the extension to the several inputs and several outputs case is straightforward, these simple cases have the advantage that we are still able to graph the problems. This will help to build some intuition for the analysis.

[3]The package is described in detail in the monograph of the package authors: Bogetoft and Otto (2001).

Table 6.9 One input–one output case: results of DEA analysis assuming constant returns to scale

	A	B	C	D	E
Input x	1.000	2.000	3.000	4.000	5.000
Output y	1.000	3.000	2.000	5.000	4.000
y/x	1.000	1.500	0.667	1.250	0.800
Efficiency	0.667	1.000	0.444	0.833	0.533

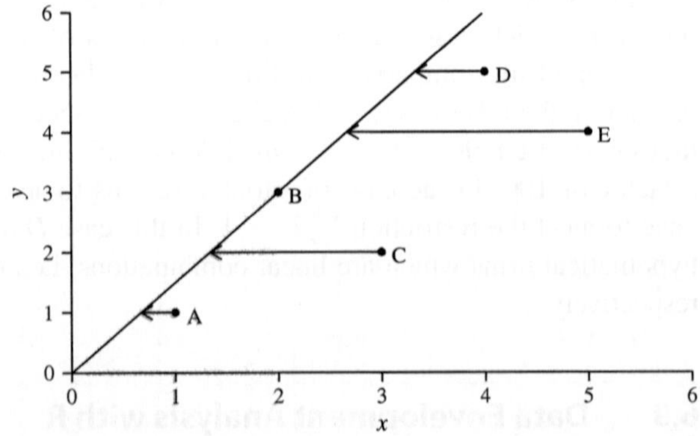

Fig. 6.7 One input–one output case

6.3.2.1 One Input–One Output

The data for the small numerical example are given in Table 6.9 and are obtained by the following code:

```
x <- matrix(1:5)
y <- matrix(c(1,3,2,5,4))
nam <- LETTERS[1:5]
eff <- y/x / max(y/x)
tab <- t(round(cbind(x,y,round(y/x,3),eff),3))
colnames(tab) <- nam
rownames(tab) <- c("input x","output y","y/x","efficiency")
```

Because of the assumption of constant returns to scale we simply calculate the ratio of output to input. We find that DMU B has the highest ratio. Dividing all ratios of output to input by the maximal ratio results in efficiency scores. We observe, e.g., that DMU C would only require 44 % of its actual input to produce its output level if it would produce with the same efficiency as DMU B which serves as the benchmark in this case. The situation is displayed in Fig. 6.7 and is obtained by:

```
library(Benchmarking)
dea.plot(x,y,RTS="crs",ORIENTATION="in-out",pch=19,cex=0.8,
         txt=LETTERS[1:length(x)],las=1)
for (i in c(1,3:5)) arrows(x[i,1],y[i,1],x[i,1]*eff[i,1],y[i,1],
                          length=0.1)
```

The arrows indicate the movement towards the efficient frontier when reducing the actual input level towards the efficient input level.

Table 6.10 Two inputs–one output case: results of DEA analysis assuming constant returns to scale

	A	B	C	D	E
x1: Input	4.00	7.00	8.00	5.00	2.00
x2: Input	3.00	2.00	1.00	5.00	5.00
y: Output	1.00	1.00	1.00	1.00	1.00

6.3.2.2 Two Inputs–One Output

The data for the two inputs–one output case example is given in Table 6.10. We set up the LP problem for firm A. The LP problem in multiplier form is noted as:

$$\begin{aligned} (LP_o) \quad & \max_{\mathbf{v},\mathbf{u}} \mathbf{u}\mathbf{y}_o \\ \text{subject to} \quad & \mathbf{v}\mathbf{x}_o = 1 \\ & -\mathbf{v}X + \mathbf{u}Y \leq 0 \\ & \mathbf{v} \geq 0, \ \mathbf{u} \geq 0. \end{aligned} \tag{6.31}$$

The inequality restrictions that for all firms output must not exceed weighted input are expressed using Matrix A1 and vector b1 in the simplex call. The standardization of the weighted input to value 1 is expressed using matrix A3 and vector b3 (equality constraint).

```
x1 <- c(4,7,8,5,2)
x2 <- c(3,2,1,5,5)
y <- rep(1,5)
library(boot)
simplex(a = c(1,0,0),
        A1 = cbind(1,-x1,-x2), b1 = rep(0,5),
        A3 = matrix(c(0,4,3),ncol=3), b3=1,
        maxi = TRUE)

##
## Linear Programming Results
##
## Call : simplex(a = c(1, 0, 0), A1 = cbind(1, -x1, -x2), b1 = rep(0,
##     5), A3 = matrix(c(0, 4, 3), ncol = 3), b3 = 1, maxi = TRUE)
##
## Maximization Problem with Objective Function Coefficients
## x1 x2 x3
##  1  0  0
##
##
## Optimal solution has the following values
##     x1     x2     x3
## 1.0000 0.1429 0.1429
## The optimal value of the objective  function is 1.
```

The solution for firm A shows that this firm is efficient (value of x1 is 1).

In the same manner this LP problem has to be set up for all n firms. We therefore use a for-loop. The first column of the matrix for collecting the results of the simplex-algorithms shows the efficiency scores for the n DMUs.

```
x1 <- c(4,7,8,5,2)
x2 <- c(3,2,1,5,5)
y <- rep(1,5)
```

```
erg  <- matrix(NA,5,3)
for (i in 1:5){
  e <- simplex(a = c(1,0,0),
        A1 = cbind(1,-x1,-x2), b1 = rep(0,5),
        A3 = matrix(c(0,x1[i],x2[i]),ncol=3), b3=1,
        maxi = TRUE)
  erg[i,] <- e$soln}
erg

##         [,1]    [,2]   [,3]
## [1,] 1.0000 0.14286 0.1429
## [2,] 0.9091 0.09091 0.1818
## [3,] 1.0000 0.10000 0.2000
## [4,] 0.7000 0.10000 0.1000
## [5,] 1.0000 0.14286 0.1429
```

With the dea() command in the Benchmarking package, we obtain the same results:

```
library(Benchmarking)
x1 <- c(4,7,8,5,2)
x2 <- c(3,2,1,5,5)
y <- rep(1,5)
X <- cbind(x1,x2)
Y <- matrix(y)
dea(X,Y,RTS="crs",ORIENTATION="in",SLACK=T)

## [1] 1.0000 0.9091 1.0000 0.7000 1.0000
```

Figure 6.8, showing the isoquant, the location of the firms in the input space, and the potential movement towards the efficient frontier, is generated using the following code:

```
library(Benchmarking)
x1 <- c(4,7,8,5,2)
x2 <- c(3,2,1,5,5)
y <- rep(1,5)
dea.plot.isoquant(x1,x2,txt=LETTERS[1:5], xlim=c(0,10),
                  pch=19,cex=0.8)
segments(0,0,x1[2]*erg[2,1],x2[2]*erg[2,1],
         lty=2,lwd=1.5)
segments(0,0,x1[4]*erg[4,1],x2[4]*erg[4,1],
         lty=2,lwd=1.5)
arrows(x1[2],x2[2],x1[2]*erg[2,1],x2[2]*erg[2,1],
       length=0.1,angle=20)
arrows(x1[4],x2[4],x1[4]*erg[4,1],x2[4]*erg[4,1],
       length=0.1,angle=20)
```

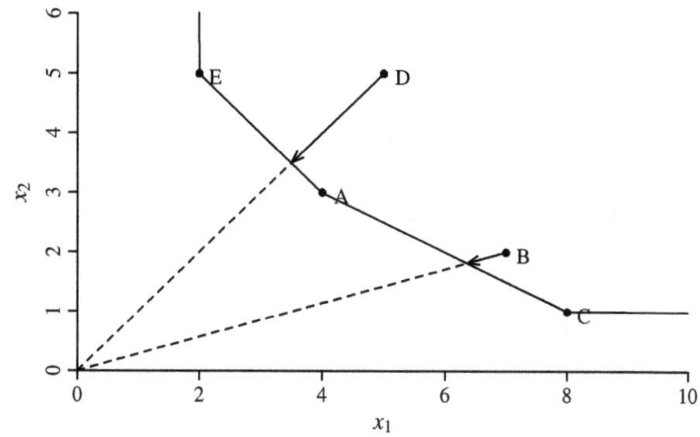

Fig. 6.8 Two inputs–one output case

We also solve the LP problem using its dual that is the envelopment form.

$$\begin{aligned} &(DLP_o) && \min_{\theta,\boldsymbol{\lambda}} \theta \\ &\text{subject to} && \theta \mathbf{x}_o - X\boldsymbol{\lambda} \geq 0 \\ &&& Y\boldsymbol{\lambda} \geq \mathbf{y}_o \\ &&& \boldsymbol{\lambda} \geq 0. \end{aligned} \tag{6.32}$$

As an example, we choose the LP for DMU B:

```
library(boot)
x1 <- c(4,7,8,5,2)
x2 <- c(3,2,1,5,5)
y <- rep(1,5)
simplex(a = c(1,0,0,0,0,0),
  A2 = cbind(c(0,7,2),rbind(1,-x1,-x2)), b2 = c(1,0,0),
  A1 = matrix(-c(0,1,1,1,1,1),ncol=6), b1=0)

##
## Linear Programming Results
##
## Call : simplex(a = c(1, 0, 0, 0, 0, 0), A1 = matrix(-c(0, 1, 1, 1, 1,
##     1), ncol = 6), b1 = 0, A2 = cbind(c(0, 7, 2), rbind(1, -x1,
##     -x2)), b2 = c(1, 0, 0))
##
## Minimization Problem with Objective Function Coefficients
## x1 x2 x3 x4 x5 x6
##  1  0  0  0  0  0
##
##
## Optimal solution has the following values
##     x1     x2     x3     x4     x5     x6
## 0.9091 0.4091 0.0000 0.5909 0.0000 0.0000
## The optimal value of the objective  function is 0.909090909090909.
```

In the simplex call a is the vector specifying the target function to be minimized. As we want to minimize θ, only the first element is set to 1. The remaining five elements for the λ-coefficients are set to 0. Matrix $A1$ and vector $b1$ are used to define a $\leq$ restriction to ensure all λ-coefficients are $\geq$ 0. Matrix $A2$ and vector $b2$ are set to meet the $\geq$ restrictions. The first row states the restriction that weighted output of the n firms is at least the actual output. The second and third row ensure that weighted input of all n firms is less than actual input times the radial potential input reduction (θ). We observe that we obtain the same result for potential radial input reduction as in the LP in multiplier form.

6.3.2.3 One Input–Two Outputs

The data for the efficiency analysis with one input and two outputs are given in Table 6.11. Note that input is already normalized to 1. The LP for the firms are quite similar to the case just discussed and we only have to make some minor adjustments. We first set up the LP and use the function `simplex()` to solve the tableau for firm A and B.

We now choose prices for outputs y_1 and y_2 to maximize sales. The restrictions we have to meet are the normalization that weighted input is 1 (matrix A3 and vector b3), that weighted output does not exceed weighted input (A1 and b1) and the non-negativity constraints for output prices (A2 and b2). The solution is obtained by:

```
y1 <- c(1,2,5,4,8)
y2 <- c(8,5,6,2,1)
x <- rep(1,5)
library(boot)
simplex(a = c(0,1,8),
        A1 = cbind(-1,y1,y2), b1 = rep(0,5),
        A3 = matrix(c(1,0,0),ncol=3), b3=1,
        A2 = rbind(c(0,1,0),c(0,0,1)),b2=c(0,0),
        maxi = TRUE)

##
## Linear Programming Results
##
## Call : simplex(a = c(0, 1, 8), A1 = cbind(-1, y1, y2), b1 = rep(0, 5),
##     A2 = rbind(c(0, 1, 0), c(0, 0, 1)), b2 = c(0, 0), A3 = matrix(c(1,
##         0, 0), ncol = 3), b3 = 1, maxi = TRUE)
##
## Maximization Problem with Objective Function Coefficients
## x1 x2 x3
##  0  1  8
##
##
## Optimal solution has the following values
##    x1    x2    x3
## 1.000 0.000 0.125
## The optimal value of the objective  function is 1.
```

We observe that firm A is efficient (value of objective function). The LP for firm B is almost identical, except that we use the different output quantities in the

Table 6.11 One input–two outputs case: results of DEA analysis assuming constant returns to scale

	A	B	C	D	E
x: Input	1.000	1.000	1.000	1.000	1.000
y1: Output	1.000	2.000	5.000	4.000	8.000
y2: Output	8.000	5.000	6.000	2.000	1.000
Efficiency	1.000	0.706	1.000	0.605	1.000

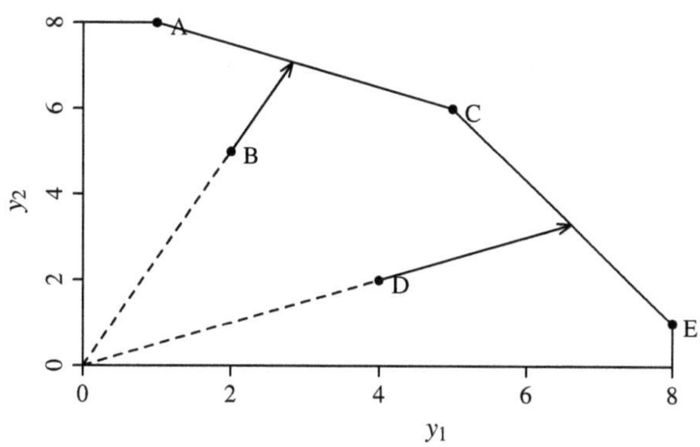

Fig. 6.9 One input–two outputs case

specification of the parameters of the objective function (`a = c(0,2,5)`). We then obtain 0.7059 as efficiency score for firm B. That means firm B is inefficient and could potentially either reduce its input by almost 30 % or increase its output radially by the factor 1/0.7059=1.41663.

Using the `dea()` command from the Benchmarking package, we obtain the results for all n DMUs. Using the option `RTS="crs"` (RTS for returns to scale) we define constant returns to scale (crs). To illustrate the LP problem, we use the plot facilities of the Benchmarking package and indicate the potential radial increase in output by arrows (see Fig. 6.9).

```
library(Benchmarking)
y1 <- c(1,2,5,4,8)
y2 <- c(8,5,6,2,1)
x <- rep(1,5)
Y <- cbind(y1,y2)
X <- matrix(x)
e <- dea(X,Y,RTS="crs",ORIENTATION="in",SLACK=T)$eff
#plot
dea.plot.transform(y1,y2,txt=LETTERS[1:5],pch=19,cex=0.8)
segments(0,0,y1[2],y2[2],lty=2,lwd=1.5)
segments(0,0,y1[4],y2[4],lty=2,lwd=1.5)
segments(y1[3],y2[3],y1[5],y2[5],lty=1)
arrows(y1[2],y2[2],y1[2]/e[2],y2[2]/e[2],
       length=0.1,angle=20)
arrows(y1[4],y2[4],y1[4]/e[4],y2[4]/e[4],
       length=0.1,angle=20)
```

6.4 Recommended Reading

The reference text for data envelopment analysis is

- Cooper WW, Seiford LM, Tone K (2007) Data envelopment analysis. Springer, New York.

Despite its coverage of many specialized aspects of DEA, the book is very didactical and starts with most basic numerical examples. A less didactical and more concise overview is

- Cooper WW, Seiford LM, Zhu J (2011) Handbook on data envelopment analysis, chap Data envelopment analysis: history, models and interpretations. Springer, New York, pp 1–39.

An overview of DEA is also provided (chapters 6 and 7) by

- Coelli TJ, Prasada Rao DS, O'Donnell CJ, Battese GE (2005) An introduction to efficiency and productivity analysis, 2nd edn. Springer, New York
- Thanassoulis E, Portela MCS, Despic O (2008) The measurement of productive efficiency and productivity growth, chap Data envelopment analysis: the mathematical programming approach to efficiency analysis. Oxford University Press, Oxford.

Chapters 4 and 5 of

- Bogetoft P, Otto L (2001) Benchmarking with DEA, SFA and R. Springer, New York.

are devoted towards deterministic data envelopment analysis and provide examples using their R-package Benchmarking.

6.5 Exercises

Consider the following input and output data for 8 DMUs (Table 6.12):

1. Display the LP problem graphically using the function `dea.plot.transform()` and explain the plot. Which firms are inefficient? Estimate the inefficiency scores by eyesight.
2. Consider the LP in multiplier form and set up the problem for DMU A.
3. Use the function `simplex()` of the boot-package to find the solution of the LP for DMU A.
4. Solve the LP in multiplier form for DMU C.
5. State the dual of the LP in multiplier form that is the LP in envelopment form for DMU C.
6. Solve the LP in envelopment form for DMU C.
7. Solve the LP using the function `dea()` from the Benchmarking package. Store the returned object and study its contents (use here, e.g., `names()` and `summary()`).

Table 6.12 Example

	A	B	C	D	E	F	G	H
x: Input	1.00	1.00	1.00	1.00	1.00	1.00	1.00	1.00
y1: Output	1.00	3.00	2.00	5.00	4.00	8.00	6.00	4.00
y2: Output	8.00	8.00	5.00	6.00	2.00	1.00	2.00	4.00

References

Bogetoft P, Otto L (2001) Benchmarking with DEA, SFA and R. Springer, New York
Coelli TJ, Prasada Rao DS, O'Donnell CJ, Battese GE (2005) An introduction to efficiency and productivity analysis, 2nd edn. Springer, New York
Cooper WW, Seiford LM, Tone K (2007) Data envelopment analysis. Springer, New York
Cooper WW, Seiford LM, Zhu J (2011) Handbook on data envelopment analysis, chap Data envelopment analysis: history, models and interpretations. Springer, New York, pp 1–39
Thanassoulis E, Portela MCS, Despic O (2008) The measurement of productive efficiency and productivity growth, chap Data envelopment analysis: the mathematical programming approach to efficiency analysis. Oxford University Press, Oxford

Stochastic Data Envelopment Analysis

7

Abstract

The deterministic data envelopment analysis is descriptive. The observed firms are compared with a (most often synthetic) benchmark firm which is either observed or constructed as a linear combination of observed firms. The stochastic data envelopment analysis now enriches the analysis by speculating about firms which have not been observed. The basic idea is that there would probably be an even more efficient benchmark firm if only more firms would have been observed. According to this speculation, the deterministic DEA efficiency scores are probably too optimistic. In this chapter, we discuss how stochastic considerations can be added to data envelopment analysis.

7.1 Introduction

The data envelopment analysis (DEA) is known as a deterministic method. The deviation from an individual firm from the efficient frontier is taken completely as due to inefficient production. There is no stochastic disturbance regarded which could have caused the deviation from the efficient frontier despite efficient production.

DEA is a rather intuitive and practical method. Having observed an amount of n firms, one picks the firms belonging to the efficient set and combines them linearly towards a synthetic firm, which then serves as a benchmark for the firm k under analysis. Probably, the fact that no dubious statistical assumptions (e.g., invented probability distributions) are necessary makes the DEA so popular among practitioners. Nevertheless, statisticians felt rather uncomfortable with the deterministic concept, missing any probability statements.

This putative gap has been closed within the last 15 years with the development of the so-called Stochastic Data Envelopment Analysis (SDEA). In this chapter, we will have a closer look at the SDEA. As the main tool used in SDEA is the bootstrap

A. Behr, *Production and Efficiency Analysis with R*,
DOI 10.1007/978-3-319-20502-1_7

method, we first give a brief introduction to the bootstrap before discussing its use in SDEA.

7.2 Population and Simple Random Samples

We first define two vectors to examine a numerical example ($N = 10$).

```
x <- c(2, 4, 5, 7, 9, 11, 14, 16, 17, 18)
y <- c(2, 4, 14, 11, 19, 15, 14, 10, 20, 16)
N <- length(x)
```

Now suppose we want to draw a simple random sample S of size $n = 6$ from the set $\mathscr{S}$ of possible random samples. For the moment, we focus on the variable Y only. What can we say about the mean of the random sample? We know that the population mean is the expected value of the sample mean, as the sample mean is an unbiased estimation function for the population mean, given as

$$\mathrm{E}(\hat{\bar{y}}) = \mathrm{E}\left(\frac{1}{n}\sum_{s} y_k\right) = \frac{1}{n}\sum_{s}\mathrm{E}(y_k) = \frac{1}{n}\sum_{s}\bar{y} = \frac{1}{n}n\bar{y} = \bar{y}. \tag{7.1}$$

The variance of the sample mean is

$$\mathrm{Var}(\hat{\bar{y}}) = \frac{\left(1-\frac{n}{N}\right)}{n}S_y^2 \quad \text{with} \quad S_y^2 = \frac{1}{N-1}\sum_{U}(y_k-\bar{y})^2. \tag{7.2}$$

For the numerical example we find

```
n <- 6
m.y <- mean(y);m.y
## [1] 12.5
v.m.y <- (1-n/N)/n*var(y);v.m.y
## [1] 2.315
```

Now assume that we would not know the analytic expression of the variance of the sample mean (the derivation in fact is somewhat tedious). One way to obtain the variance is complete enumeration of $\mathscr{S}$ which contains all possible samples of size n drawn from the population of size N. The number M of possible samples is obtained using the binomial coefficient N over n as we are interested in all combinations without replacement (that is not regarding the ordering of sample). For each of the M samples, we can calculate the sample mean to obtain the distribution of the sample mean and to calculate its variance.

```
# setup
M <- choose(N,n);M

## [1] 210

sp.index <- combn(N,n)
sp <- matrix(y[sp.index],ncol = M,byrow=F)
mu <- apply(sp,2,mean)
tab0 <- table(mu)/length(mu)

# estimate variance
var(mu)*(M-1)/M

## [1] 2.315

v.m.y # analytic

## [1] 2.315

# lowest and highest deviation from var(y)
va <- apply(sp,2,var)
no1 <- which(abs(va-var(y)) == min(abs(va-var(y))))[1];no1

## [1] 79

no2 <- which(abs(mu-mean(y)) == max(abs(mu-mean(y))))[1];no2

## [1] 200
```

The exact distribution of the arithmetic mean can be approximated by sampling repeatedly without replacement. The larger the number of samples drawn, the better the approximation will be. Here, we draw $B = 1000$ samples.

```
B <- 1000
v <- rep(NA,B)
set.seed(123)
for (i in 1:B) v[i] <- mean(sample(y,n))
tab1 <- table(v)/B

# estimate variance
var(v)*(B-1)/B

## [1] 2.359

v.m.y # analytic

## [1] 2.315
```

Graphically we can compare the exact and the approximated distribution (see Fig. 7.1).

```
# exact
plot(tab0,type="h",xlab="Sample mean",
     ylab="Rel. frequency",axes=F,
     col="darkgrey")
```

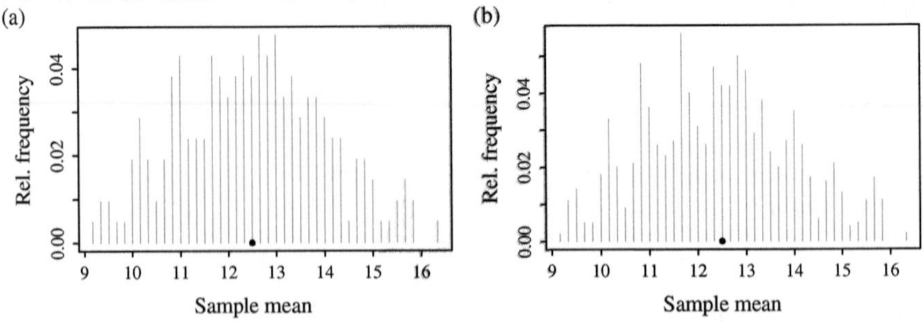

Fig. 7.1 Distribution of sample mean. The *black dot* represents the sample mean of the original y-vector. (**a**) Exact. (**b**) Approximated

```
points(m.y,0,pch=19)
axis(1)
axis(2)
# approximated
plot(tab1,type="h",xlab="Sample mean",
     ylab="Rel. frequency",axes=F,
     col="darkgrey")
points(m.y,0,pch=19)
axis(1)
axis(2)
```

7.3 The Bootstrap

So far, we have examined all possible samples contained in $\mathscr{S}$. In practical problems only one specific sample $S = s$ has been drawn and the remaining $N - n$ elements of the population are unknown. Therefore, the only information available is the drawn sample. The population mean and the distribution and variance of the sample mean are unknown and have to be estimated. One way is to use analytic estimation functions. Let us for the moment assume we have drawn sample `no1` (which is sample number 79 of the 210 samples and has the lowest deviation):

```
ys <- sp[,no1];ys

## [1]  2 14 11 19 10 16
```

Our estimator of the variance of the sample mean is

$$\widehat{\mathrm{Var}}(\hat{\bar{y}}) = \frac{\left(1 - \frac{n}{N}\right)}{n}\hat{S}_y^2, \qquad \text{with} \qquad \hat{S}_y^2 = \frac{1}{n-1}\sum_s (y_k - \bar{y}_s)^2 . \tag{7.3}$$

```
mean(ys)

## [1] 12

(1-n/N)/n*var(ys)

## [1] 2.32

m.y  # true mean

## [1] 12.5

v.m.y  # true variance

## [1] 2.315
```

We slightly underestimate the population mean and come very close to the true variance of the sample mean.

An alternative way to obtain an estimate of variance of the sample mean is the bootstrap method.[1] The idea is that the sample relates to the population in a similar way as a sub-sample of the sample relates to the sample. From the original sample s, we sample with replacement B sub-samples (bootstrap-samples) s^* and use the variance of the means of the bootstrap samples as an estimator for the variance of the sample mean drawn from the population. Because we sample from a finite population, we consider a correction factor (one minus sampling fraction) in all variance estimates.

```
B <- 1000
e <- rep(NA,B)
set.seed(7)
for (i in 1:B) e[i] <- mean(sample(ys,n,replace=T))
mean(e)

## [1] 11.91

mean(ys)

## [1] 12

# bootstrap estimator of variance of sample mean:
(1-n/N)*var(e)

## [1] 1.959

# analytic estimator of variance of sample mean:
(1-n/N)/n*var(ys)

## [1] 2.32
```

[1]In Wikipedia, we find the following explanation: "Widely attributed to The Surprising Adventures of Baron Munchausen, (1781) by Rudolf Erich Raspe, where the eponymous Baron pulls himself out of a swamp by his hair (specifically, his pigtail), though not by his bootstraps; misattribution dates to US, 1901".

```
v.m.y  # true variance of sample mean

## [1] 2.315
```

Note that in this very small example, there is a noteworthy difference between the analytic and the bootstrap estimators of variance of the sample mean.

We consider another sample s no2 (which is sample number 200 of the 210 samples and had the highest deviation) from $\mathscr{S}$:

```
ys <- sp[,no2];ys

## [1] 14 19 15 14 20 16

mean(ys)

## [1] 16.33

# analytic estimator of variance of sample mean:
(1-n/N)/n*var(ys)

## [1] 0.4444

m.y  # true mean

## [1] 12.5

v.m.y  # true variance of sample mean

## [1] 2.315
```

Note that based on this sample, we overestimate the population mean by about 30.67 % and underestimate that variance of the sample mean by about −80.8 %. As the bootstrap samples rely completely on the original sample, the deviations between sample and population parameters of course cannot be cured by the bootstrap.

```
B <- 1000
e <- rep(NA,B)
set.seed(7)
for (i in 1:B) e[i] <- mean(sample(ys,n,replace=T))
 # bootstrap estimator of variance of sample mean
n/(n-1)*(1-n/N)*var(e)

## [1] 0.4492
```

Note that we use the factor $n/(n-1)$ to take into account the loss of one degree of freedom and to obtain an unbiased estimate of variance of the sample mean. We find that the result of the bootstrap procedure is almost identical to the analytic result based on the specific sample. Hence, the bootstrap is an alternative method for obtaining, e.g., variance estimates which might be very welcome if analytic expressions are difficult to derive, but, of course, the bootstrap cannot remedy bad luck in sampling.

7.4 Sampling and DEA

We first look at the population data and carry out a DEA assuming variable returns to scale.

```
library(Benchmarking)
# data
x <- c(2, 4, 5, 7, 9, 11, 14, 16, 17, 18)
y <- c(2, 4, 14, 11, 19, 15, 14, 10, 20, 16)
t(data.frame(x,y, row.names=LETTERS[1:10]))

##   A B  C  D  E  F  G  H  I  J
## x 2 4  5  7  9 11 14 16 17 18
## y 2 4 14 11 19 15 14 10 20 16

# DEA (population)
d <- round(dea(x,y,RTS=1,ORIENTATION="in")$eff,3)
names(d) <- LETTERS[1:10];d

##     A     B     C     D     E     F     G     H     I
## 1.000 0.625 1.000 0.607 1.000 0.527 0.357 0.250 1.000
##     J
## 0.367
```

For example for firm H, we obtain an efficiency value of 0.25. Therefore, this unit could either reduce its inputs by about 75 % or increase its output by about 100 % (see Fig. 7.2) by producing efficiently.

But now suppose that a sample of $n = 6$ has been drawn and a DEA is carried out based on the sample. As an example, we look at a sample containing the firms A, B, D, H, I, J (corresponding to the identification numbers 1, 2, 4, 8, 9, 10).

```
# create sample
set.seed(7)
s <- sort(sample(1:10,n));s

## [1]  1  2  4  8  9 10
```

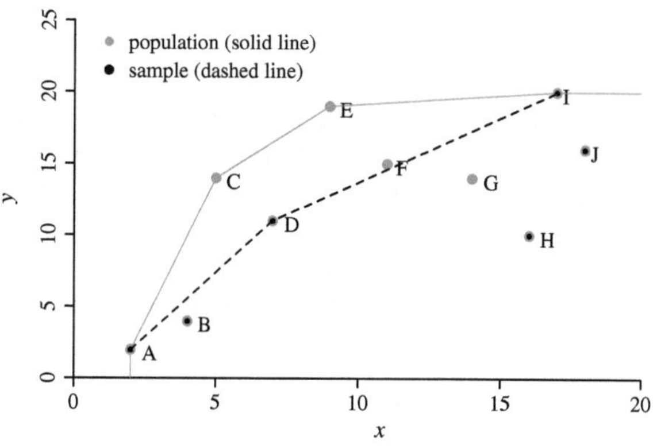

Fig. 7.2 Frontier and observations, population (10 observations) and sample (6 out of 10 observations)

```
# obtain data of the sample
xs <- x[s]
ys <- y[s]
# DEA (sample)
ds <- round(dea(xs,ys,RTS=1,ORIENTATION="in")$eff,3)
names(ds) <- LETTERS[s];ds

##     A     B     D     H     I     J
## 1.000 0.778 1.000 0.403 1.000 0.698
```

The comparison between population and sample shows:

```
rbind(population=d[s],sample=ds)

##            A     B     D     H I     J
## population 1 0.625 0.607 0.250 1 0.367
## sample     1 0.778 1.000 0.403 1 0.698
```

Both analyses can be illustrated graphically using the following code (see Fig. 7.2).

```
dea.plot(x,y,txt=LETTERS[1:10],xlab='x',ylab='y',
         cex=1.2,pch=19,col="darkgrey",ylim=c(0,25))
points(x[s],y[s],pch=19,col=1,cex=0.5)
lines(x[s][ds==1],y[s][ds==1],lty=2)
legend(0.5,25,c("population (solid line)",
                "sample (dashed line)"),
       pch=c(19),col=c("darkgrey","black"),bty="n")
```

In the sample, firm D belongs to the efficient set and therefore has efficiency 1. Obviously, the result of the efficiency calculation depends on the sample S which is drawn from the space $\mathscr{S}$.

```
withD <- apply(sp.index,2,function(z) 4%in%z)
sum(withD)

## [1] 126

mean(withD)

## [1] 0.6
```

In our example, only the fraction $n/N = 6/10 = 0.6$ of all $M = 210$ samples contain firm D. We calculate the resulting efficiency score for firm D (θ_D) in the 126 samples that include firm D. Note that because it is random which sample S will be drawn, the efficiency score is random depending also on S: $\theta_D(S)$. We derive the probability distribution of the efficiency score of firm D. As we only regard simple random sampling, all M samples have the same probability to occur.

```
# take all samples (126) which include firm D
sp.index4 <- sp.index[,withD]
# input/output data for these samples
```

```
sp4.y <- matrix(y[sp.index4],ncol = M*n/N,byrow=F)
sp4.x <- matrix(x[sp.index4],ncol = M*n/N,byrow=F)
# DEA for the first sample which includes firm D
dea(matrix(sp4.x[,1]),matrix(sp4.y[,1]))

## [1] 1.0000 0.6250 1.0000 0.6071 1.0000 0.5273

# DEA for each sample which includes firm D
M4 <- (M*n/N)
e4 <- rep(NA,M4)
for (i in 1:M4){
 e4[i] <- dea(matrix(sp4.x[,i]),
          matrix(sp4.y[,i]))$eff[sp.index4[,i]==4]
}
```

The object e4 holds all 126 efficiency scores for firm D from each sample. The results can be plotted with the following code (see Fig. 7.3).

```
plot(table(round(e4,2))/M4,lwd=1,
     xlab="Efficiency scores for firm D",
     ylab="Probability")
```

If a firm belongs to the efficient set in the population, it will belong to the efficient set in all samples containing this firm. We derive the distribution of efficiency scores for the firms not belonging to efficient set in the population by examining all relevant samples.

```
Mk <- M*n/N # number of samples including element k
ie <- (1:N)[d!=1] # inefficient firms
nie <- length(ie) # number of inefficient firms
ee <- matrix(NA,Mk,nie)
for (j in 1:nie){
  k <- ie[j]
  withk <- apply(sp.index,2,function(z) k%in%z)
  sp.indexk <- sp.index[,withk]
```

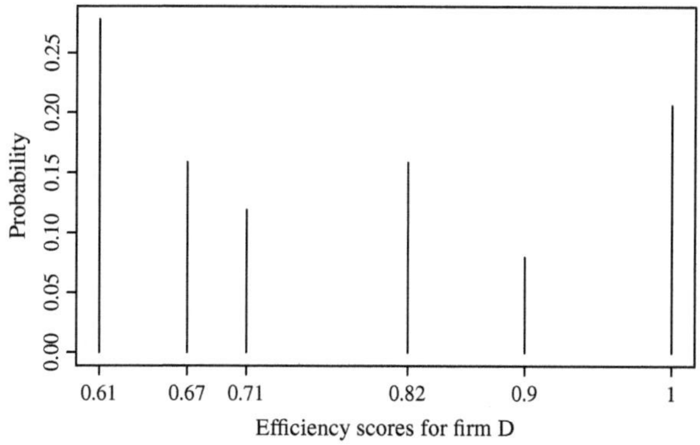

Fig. 7.3 Distribution of efficiency scores of firm D

```
  spk.y <- matrix(y[sp.indexk],ncol = Mk,byrow=F)
  spk.x <- matrix(x[sp.indexk],ncol = Mk,byrow=F)
  for (i in 1:Mk) ee[i,j] <- dea(matrix(spk.x[,i]),
                    matrix(spk.y[,i]))$eff[sp.indexk[,i]==k]
}
```

The results can be plotted with the following code (see Fig. 7.4).

```
for (i in 1:nie){
  plot(table(round(ee[,i],2))/Mk,lwd=1,
       xlab=paste("Efficiency scores, firm", LETTERS[ie][i]),
       ylab="Probability")
}
```

We find the following intervals (min($\theta(S)$) and max($\theta(S)$)) means E($\theta(S)$) and biases $\theta(S) - \theta$ of efficiency scores for the six inefficient firms:

```
eet <- data.frame(round(rbind(
        apply(ee,2,min),
        apply(ee,2,max),
        apply(ee,2,mean),
        d[ie],
        apply(ee,2,mean)-d[ie]),3))
names(eet) <- LETTERS[ie]
rownames(eet) <- c("min","max","mean","pop. efficiency","bias")
eet
```

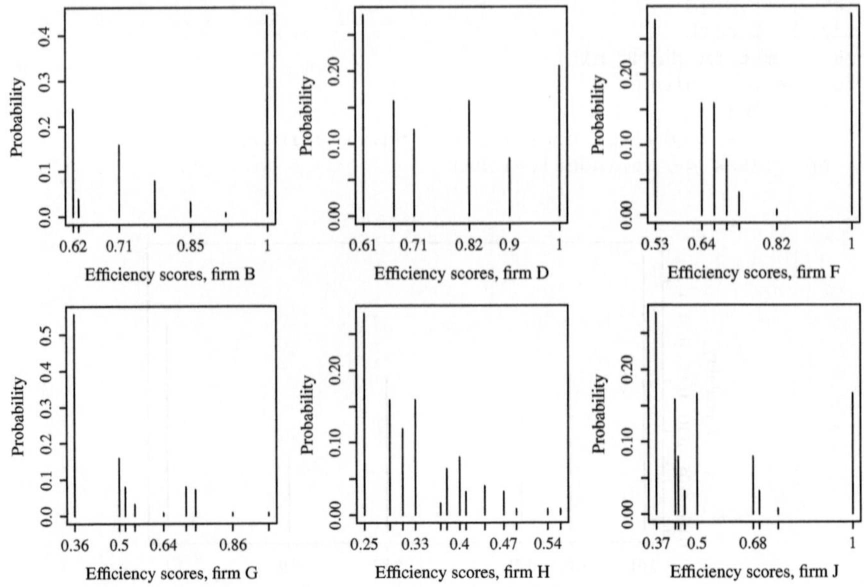

Fig. 7.4 Distribution of efficiency scores for firms not in the efficient set

```
##                       B     D     F     G     H     J
## min                0.625 0.607 0.527 0.357 0.250 0.367
## max                1.000 1.000 1.000 0.971 0.562 1.000
## mean               0.826 0.768 0.724 0.465 0.325 0.552
## pop. efficiency 0.625 0.607 0.527 0.357 0.250 0.367
## bias               0.201 0.161 0.197 0.108 0.075 0.185
```

7.5 DEA and the Naive Bootstrap

In the last section, we have examined the probability distributions of the efficiency scores. In real world problems, of course, the population cannot be completely enumerated as one is stuck with the single sample that has been drawn. We now want to ask whether any insights can be gained from SDEA beyond the classical non-stochastic DEA using bootstrap methods.

Applying the idea of bootstrapping directly to the drawn sample causes some problems. In the bootstrap samples, some firms of the original sample will be missing and other firms will be included twice or multiple times. In contrast to, e.g., regression analysis, in DEA it is not possible to include multiple observed firms more than once, as the linear program would be identical. Therefore, we can either use only the unique firms in the bootstrap sample which makes the number of unique firms m ($m <= n$) in the bootstrap sample random, or we sub-sample without replacement with $m < n$.

7.5.1 Some Notation

We denote a specific sample with subscript s and a bootstrap sample with superscript b. To a specific firm, we refer to with subscript k. θ_k denotes the efficiency score for firm k in the population. $\theta_{s,k}$ denotes the efficiency of firm k in sample s and $\theta^b_{s,k}$ denotes it in the bootstrap sample b drawn from sample s. M denotes the number of all possible samples

$$M = |\mathscr{S}| = \binom{N}{n} \tag{7.4}$$

$I_{(s \ni k)}$ is the inclusion indicator. It takes value 1 if element k is contained in the sample s and 0 otherwise. M_k is the number of all samples containing firm k

$$M_k = \sum_{s=1}^{M} I_{(s \ni k)} = M\frac{n}{N}. \tag{7.5}$$

We use the expression $s \ni k$ to qualify the samples that contain firm k. The bias b_k, we define as the difference between the average efficiency score $\hat{\theta}_k$ obtained for firm

k in all M_k relevant samples and its efficiency score in the population θ_k

$$b_k = \bar{\theta}_k - \theta_k = \frac{1}{M_k} \sum_{s \ni k} (\theta_{sk} - \theta_k) \, . \tag{7.6}$$

The estimate of the bias based on sample s obtained through the bootstrap is

$$\hat{b}_{sk} = \frac{1}{B^*} \sum_{b \ni k} \left(\theta^b_{sk} - \theta_{sk}\right), \qquad \text{with} \qquad B^* = \sum_{b=1}^{B} I_{(b \ni k)}. \tag{7.7}$$

Note that while we draw B bootstrap samples, only B^* bootstrap samples will contain firm k. We denote this by $b \ni k$. An estimate of the bias based on all M_k samples containing k is then

$$\hat{b}_k = \frac{1}{M_k} \sum_{s \ni k} \hat{b}_{sk} = \frac{1}{M_k} \sum_{s \ni k} \left(\frac{1}{B^*} \sum_{b \ni k} \left(\theta^b_{sk} - \theta_{sk}\right) \right). \tag{7.8}$$

Note that b_k and $\hat{b}_k$ can only be calculated in simulation studies as they rely on observing the population. If sample size n and sub-sample size m are small, it is possible to completely enumerate all B^e possible sub-samples where

$$B^e = \binom{n}{m}. \tag{7.9}$$

In the case of simple random sampling, the number of sub-samples containing firm k is denoted by B^e_k and obtained by

$$B^e_k = \binom{n-1}{m-1}. \tag{7.10}$$

7.5.2 Bootstrap Estimation Based on a Specific Sample

The procedure can be characterized by the following steps:

1. Obtain a bootstrap sample $\{x, y\}_b$ of size m from the original sample $\{x, y\}_s = \{(x_1, y_1), \cdots, (x_k, y_k), \cdots, (x_n, y_n), \}_s$ through sampling without replacement.
2. Calculate DEA efficiency scores for the m units in the bootstrap sample to obtain bootstrap efficiency estimates θ^b_k.
3. Repeat the procedure B-times and evaluate the distribution of the B^*-bootstrap estimates $\hat{\theta}^b_k$ for each k.
4. Estimate the bias b_k as the difference between $\hat{\theta}^b_k = 1/B^* \sum_{b=1}^{B^*} \hat{\theta}^b_k$ and θ_{sk}.

We use the same data as above:

```
# data
x <- c(2, 4, 5, 7, 9, 11, 14, 16, 17, 18)
y <- c(2, 4, 14, 11, 19, 15, 14, 10, 20, 16)
# create sample (like above)
n <- 6
set.seed(7)
s <- sort(sample(1:10,n));s

## [1]  1  2  4  8  9 10

# input/output data of the sample
xs <- x[s]
ys <- y[s]
# DEA
ds <- Benchmarking::dea(xs,ys,RTS=1,ORIENTATION="in")
es <- ds$eff
names(es) <- LETTERS[s]
# obtain data of the sample
xs <- x[s]
ys <- y[s]
```

We sub-sample without replacement $m = 4$ firms from the $n = 6$ firms in the sample. We first demonstrate the procedure using a single bootstrap sample (`si`)[2]:

```
# bootstrap sample
m <- 4
si <- c(1,2,3,6)
# input/ouput data of the bootstrap sample
xi <- xs[si];xi

## [1]  2  4  7 18

yi <- ys[si];yi

## [1]  2  4 11 16

# DEA of the bootstrap sample
di <- Benchmarking::dea(xi,yi,RTS=1,ORIENTATION="in")$eff
names(di) <- LETTERS[s][si]
```

The results can be plotted with the following code (see Fig. 7.5).

```
dea.plot(x,y,txt=LETTERS[1:10],cex=1.2,pch=19,col="darkgrey",
         xlab='x',ylab='y',ylim=c(0,25))
points(x[s],y[s],pch=19,col=1,cex=0.5)
lines(x[s][ds$eff==1],y[s][ds$eff==1],lty=2)
points(xi,yi,pch=1,cex=1.3)
lines(xi[di==1],yi[di==1],lty=3,lwd=1)
```

[2]Because the function `dea()` is contained in several packages we use the command `Benchmarking::dea()` for using the one implemented in the Benchmarking package.

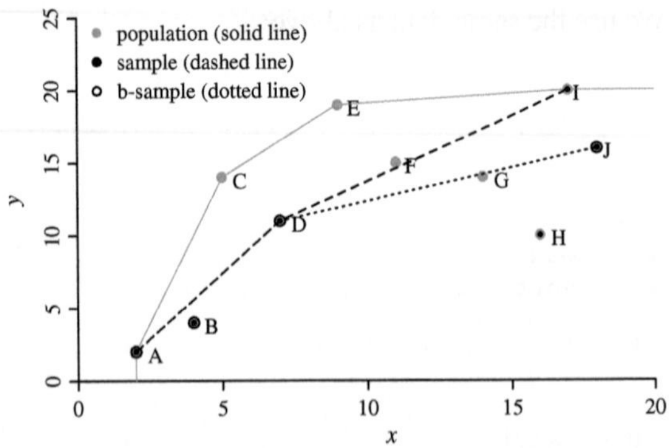

Fig. 7.5 Frontier and observations, population (10 observations), sample (6 out of 10 observations), and bootstrap-sample (4 out of 6 observations)

```
legend(0.5,25,c("population (solid line)",
                "sample (dashed line)",
                "b-sample (dotted line)"),pch=c(19,19,1),
       col=c("darkgrey","black", "black"),bty="n")
```

In this bootstrap sample, firm B produces inefficiently. Despite the fact that in this small example

$$B^e = \binom{6}{4} = 15$$

we do not enumerate completely but bootstrap as we would do with larger samples. We calculate for $B = 1000$ bootstrap samples the efficiency scores and evaluate the result.

```
m <- 4
B <- 1000
Be <- matrix(NA,B,n)
set.seed(1)
# Bootstrap
for (i in 1:B){
  si <- sample(1:n,m,replace=F)
  xi <- xs[si]
  yi <- ys[si]
  di <- Benchmarking::dea(xi,yi,RTS=1,ORIENTATION="in")
  Be[i,si] <- di$eff
}
m.Be <- apply(Be,2,function(z) mean(z,na.rm=T))
```

We use the difference between average bootstrap efficiency and efficiency scores of the original sample to estimate the bias and bias-corrected efficiency scores:

```
bias <- m.Be-es
bcor <- es-bias
round(rbind(es,m.Be,bias,bcor,pop=d[s]),3)
```

```
##         A     B     D     H I     J
## es     1 0.778 1.000 0.403 1 0.698
## m.Be   1 0.914 1.000 0.482 1 0.840
## bias   0 0.136 0.000 0.079 0 0.142
## bcor   1 0.642 1.000 0.324 1 0.555
## pop    1 0.625 0.607 0.250 1 0.367
```

It holds that an efficient firm in the population will be efficient in any sample. It holds too that an efficient firm in an original sample will be efficient in any bootstrap sample. Therefore, no bootstrap bias correction can cure a biased efficiency if in the original sample $\theta_{sk} = 1$ has been obtained.

7.5.3 Bootstrap Estimation Based on All Samples

As we have picked one specific sample to exemplify the bootstrap approach, we now examine the bootstrap results (with $B = 100$ replications) for each of the $|\mathcal{S}| = M = 210$ possible samples.

```
B <- 100
Ep <- matrix(NA,M,N)
Es <- Ep
Eb <- Ep
set.seed(1)
for (j in 1:M){
s <- sp.index[,j]
Ep[j,s] <- d[s]
Es[j,s] <- dea(x[s],y[s],RTS=1,ORIENTATION="in",SLACK=F)$eff
Be <- matrix(NA,B,N)
for (i in 1:B){
si <- sample(s,m,replace=F);si
xi <- x[si];xi
yi <- y[si];yi
Be[i,si] <- dea(xi,yi,RTS=1,ORIENTATION="in",SLACK=F)$eff
}
Eb[j,s] <- apply(Be,2,function(z) mean(z,na.rm=T))[s]
}
tab <- as.data.frame(cbind(d,apply(Es,2,
                                   function(z) mean(z,na.rm=T)),
             apply(Eb,2,function(z) mean(z,na.rm=T))))
tab <- tab[1:10,]
rownames(tab) <- LETTERS[1:10]
tab$bias <- tab[,2]-tab[,1]
tab$estb <- tab[,3]-tab[,2]
tab$diffb <- tab$estb-tab$bias
names(tab) <- c("pop.","sample","boot","bias",
                "est.bias","d.bias")
tab <- round(tab,3)
```

The results are displayed in Table 7.1. We observe that on average the approximation of the bias is quite good using the naive bootstrap approach with sub-samples

Table 7.1 Results of bootstrap estimation based on all samples

	pop.	sample	boot	bias	est.bias	d.bias
A	1.00	1.00	1.00	0.00	0.00	0.00
B	0.62	0.83	0.92	0.20	0.10	−0.10
C	1.00	1.00	1.00	0.00	0.00	0.00
D	0.61	0.77	0.88	0.16	0.11	−0.05
E	1.00	1.00	1.00	0.00	0.00	0.00
F	0.53	0.72	0.87	0.20	0.14	−0.06
G	0.36	0.47	0.61	0.11	0.14	0.03
H	0.25	0.33	0.42	0.07	0.09	0.02
I	1.00	1.00	1.00	0.00	0.00	0.00
J	0.37	0.55	0.73	0.18	0.17	−0.01

of fixed size $m < n$. Therefore, correcting the sample efficiency estimates with bootstrap estimates of the bias on average results in a considerable bias reduction.

7.6 The Simar–Wilson Approach

In the previous section, we have drawn sub-samples without replacement from the original sample to obtain bootstrap samples with m pairs x_k, y_k. This resulted in bootstrap samples which did not allow to calculate bootstrap estimates for only m of the n firms of the original sample in individual bootstrap replications. Therefore, the number of bootstrap estimates for firm k was B^* with $B^* < B$.

A completely different approach has been suggested by Simar and Wilson. Instead of sampling firms, in this approach, observed inputs of firms are randomly altered using one randomly chosen efficiency score from the efficiency scores which have been obtained for the original sample. Furthermore, the randomly chosen score is overlaid by a normal error term. This approach is not based on the idea of sampling from finite populations but rather on the ideas of superpopulation models. One implication of these approach is that even if all firms of a sector have been considered, it is still assumed that there is an overestimation bias due to neglecting "non-existing" firms. These "non-existing" firms are nevertheless regarded as "potentially" existing. In this approach, observed efficiency scores are assumed to be realizations of an existing "distribution of efficiency scores". Therefore, using a bootstrap approach shall mimic the process of realizations from this distribution.

7.6.1 Strongly Simplified Algorithm

We first examine a strongly simplified version of the approach which does not consider normally distributed disturbances to focus on the basic idea. For convenience, the efficiency parameter θ denotes the Shepard input distance (being ≥ 1) and therefore is the reciprocal of the Farrell efficiency score (which is ≤ 1).

1. Obtain a bootstrap sample of size n $\{\theta_1^*, \cdots, \theta_k^*, \cdots, \theta_n^*\}$ from the efficiency scores of the original sample $\{\hat{\theta}_1, \cdots, \hat{\theta}_k, \cdots, \hat{\theta}_n\}$ through sampling with replacement.
2. Generate fictional inputs for the bootstrap sample as $x_k^* = x_k\, \theta_k^* / \hat{\theta}_k$.
3. Calculate bootstrap DEA efficiency scores $\hat{\theta}_k^*$ using original outputs y, inputs x, and fictional bootstrap inputs x^* as the reference input for the n firms $k = 1, \ldots, n$.
4. Repeat the procedure B-times to obtain B-bootstrap estimates $\hat{\theta}_k^*$ for each k.
5. Estimate the bias $\mathrm{b}(\hat{\theta}_k) = \hat{\bar{\theta}}_k^* - \hat{\theta}_k$ with $\hat{\bar{\theta}}_k^* = 1/B \sum_{b=1}^{B} \hat{\theta}_k^*$.
6. Estimate the bias corrected (bc) efficiency score as:
   $\hat{\theta}_k^{\mathrm{bc}} = \hat{\theta}_k - \mathrm{b}(\hat{\theta}_k) = 2\hat{\theta}_k - \hat{\bar{\theta}}_k^*$.
7. Obtain a nonparametric confidence interval for the efficiency scores using quantiles (e.g., $q_{0.05}$ and $q_{0.95}$ for $\alpha = 0.1$). The quantiles have to be corrected by subtracting two times the estimated bias.

We apply this algorithm towards a small data set (5 firms) in the one input–one output case (to simplify the code, we use d for θ throughout).

```
library(Benchmarking)
x <- c(2,4,3,5,6) # input
y <- 1:5 # output
n <- length(x)
dhat <- 1/dea(x,y,RTS=1,ORIENTATION="in")$eff;dhat

## [1] 1.000 1.600 1.000 1.111 1.000

B <- 2000
Bm <- matrix(NA,B,n)
set.seed(1)
# Step 4
for (i in 1:B){
  # Step 1
  dstar <- sample(dhat,n,replace=T)
  # Step 2
  xs <- x*dstar/dhat
  # Step 3
  Bm[i,] <- 1/dea(x,y,XREF=xs,YREF=y,
                  RTS=1,ORIENTATION="in")$eff
}
# Step 5
dhat.m <- colMeans(Bm)
dhat.bias <- dhat.m-dhat
# Step 6
dhat.bc <- dhat-dhat.bias
# Step 7
ci.low <- apply(Bm,2,quantile,0.05)-2*dhat.bias
```

Table 7.2 Results of the strongly simplified algorithm

	1	2	3	4	5
dhat	1.000	1.600	1.000	1.111	1.000
dhat.m	0.934	1.560	0.942	1.079	0.903
dhat.bias	−0.066	−0.040	−0.058	−0.032	−0.097
dhat.bc	1.066	1.640	1.058	1.143	1.097
ci.low	0.853	1.422	0.915	1.061	0.818
ci.high	1.133	1.680	1.115	1.175	1.193

```
ci.high <- apply(Bm,2,quantile,0.95)-2*dhat.bias
# Results
tab <- round(rbind(dhat,dhat.m,dhat.bias,
             dhat.bc,ci.low,ci.high),3)
```

The results are displayed in Table 7.2.

7.6.2 The Simar–Wilson Algorithm

We now examine a less simplified version of the approach and add a realization of a normally distributed random variate to the bootstrapped efficiency scores. As this procedure can result in efficiency scores exceeding the value 1, we have to make an adjustment in these cases to guarantee that the randomly overlaid score has a value not exceeding 1. Furthermore, the increase in the variance of the efficiency scores caused by the addition of a random term has to be taken care of.

1. Obtain a bootstrap sample of size n $\{\theta_1^*, \cdots, \theta_k^*, \cdots, \theta_n^*\}$ from the mirrored efficiency scores of the original sample $\{\hat{\theta}_1, \cdots, \hat{\theta}_k, \cdots, \hat{\theta}_n, 2-\hat{\theta}_1, \cdots, 2-\hat{\theta}_k, \cdots, 2-\hat{\theta}_n\}$ through sampling with replacement.
2. Choose an "optimal" bandwidth h for the set of values $\{\hat{\theta}, 2-\hat{\theta}\}$.
3. Generate a realization ε of size n of a standard normal random variate and calculate $\tilde{\theta}_k = \theta_k^* + \varepsilon_k$.
4. Standardize the bootstrap efficiency scores

$$\tilde{\theta}_k^* = \bar{\hat{\theta}} + \frac{1}{\sqrt{1+h^2/\hat{\sigma}_{\hat{\theta}}^2}}(\tilde{\theta}_k - \bar{\hat{\theta}}),$$

with

$$\bar{\hat{\theta}} = \frac{1}{2n}\sum_{k=1}^{n}(\hat{\theta}_k + (2-\hat{\theta}_k)) = 1 \quad \text{and}$$

$$\hat{\sigma}_{\hat{\theta}}^2 = \frac{1}{2n-1}\sum_{k=1}^{n}[(\hat{\theta}_k - \bar{\hat{\theta}}_k)^2 + ((2-\hat{\theta}_k) - \bar{\hat{\theta}}_k)^2].$$

5. Restrict the bootstrap efficiency scores to be ≥ 1:

$$\tilde{\theta}_k^{**} = \begin{cases} \tilde{\theta}_k^*, & \text{if}\tilde{\theta}_k^* \geq 1 \\ 2 - \tilde{\theta}_k^*, & \text{otherwise.} \end{cases}$$

6. Generate fictional inputs for the bootstrap sample as $x_k^* = x_k \, \hat{\theta}_k / \tilde{\theta}_k^{**}$.
7. Calculate bootstrap DEA efficiency scores $\hat{\theta}^*$ using original outputs y and inputs x and fictional bootstrap inputs x^* as reference inputs for the n firms.
8. Repeat the procedure B-times to obtain B-bootstrap estimates $\hat{\theta}_k^*$ for each k.
9. Estimate the bias $\text{b}(\hat{\theta}_k) = \hat{\bar{\theta}}_k^* - \hat{\theta}_k$ with $\hat{\bar{\theta}}_k^* = 1/B \sum_{b=1}^{B} \hat{\theta}_k^*$.
10. Estimate the bias corrected (bc) efficiency score as: $\hat{\theta}_k^{\text{bc}} = \hat{\theta}_k - \text{b}(\hat{\theta}_k) = 2\hat{\theta}_k - \hat{\bar{\theta}}_k^*$.
11. Obtain a nonparametric confidence interval for the efficiency scores using quantiles (e.g., $q_{0.05}$ and $q_{0.95}$ for $\alpha = 0.1$). The individual estimates $\hat{\theta}_k^*$ are used to correct $\hat{\theta}$ by subtracting the estimated bias based on this bootstrap sample $\hat{\theta}_k^* - \hat{\theta}$, i.e., $\hat{\theta} - (\hat{\theta}_k^* - \hat{\theta}) = 2\hat{\theta} - \hat{\theta}_k^*$.

```
library(Benchmarking)
x <- c(2,4,3,5,6) # input
y <- 1:5 # output
n <- length(x)
# Step 1
dhat <- 1/dea(x,y,RTS=1,ORIENTATION="in")$eff
# Step 2
h <- density(c(dhat,2-dhat))$bw  # bw: bandwidth
dhat2 <- c(dhat,2-dhat)
B <- 2000
Bm <- matrix(NA,B,n)
set.seed(1)
# Step 8
for (i in 1:B){
  # Step 3
  dstar <- sample(dhat2,n,replace=T)
  eps <- rnorm(n)
  dt <- dstar+h*eps
  # Step 4
  v <- var(dhat2)
  dts <- 1+1/sqrt(1+h^2/v)*(dt-1)
  # Step 5
  dtss <- ifelse(dts > 1,dts,2-dts)
  # Step 6
  xs <- x*dtss/dhat
  # Step 7
  Bm[i,] <- 1/dea(x,y,XREF=xs,YREF=y,RTS=1)$eff
}
# Step 9
dhat.m <- colMeans(Bm)
```

```
dhat.bias <- dhat.m-dhat
# Step 10
dhat.bc <- dhat-dhat.bias

# Step 11
Bm.c <- t(apply(-Bm,1,function(z) z+2*dhat))
ci.low <- apply(Bm.c,2,quantile,0.05)
ci.high <- apply(Bm.c,2,quantile,0.95)

# Results
tab <- round(rbind(dhat,dhat.m,dhat.bias,
                   dhat.bc,ci.low,ci.high),2)
```

The results are displayed in Table 7.3.

We compare the results with the implemented bootstrap routine in the "FEAR" package. The bootstrap procedure can be invoked using the `boot.sw98()` command.

```
library(FEAR)
b <- boot.sw98(matrix(x,1),matrix(y,1),RTS=1,alpha=0.1)
tab <- round(rbind(dhat=b$dhat,dhat.m=rowMeans(b$boot),
                   bias=b$bias,dhat.bc=b$dhat.bc,
                   ci.low=b$conf.int[,1],
                   ci.high=b$conf.int[,2]),2)
```

The results obtained using `boot.sw98()` are almost identical and are displayed in Table 7.4.

Table 7.3 Results of the Simar–Wilson algorithm

	1	2	3	4	5
dhat	1.00	1.60	1.00	1.11	1.00
dhat.m	0.86	1.45	0.87	1.01	0.85
dhat.bias	−0.14	−0.15	−0.13	−0.10	−0.15
dhat.bc	1.14	1.75	1.13	1.21	1.15
ci.low	1.01	1.62	1.01	1.12	1.01
ci.high	1.32	1.94	1.27	1.36	1.39

Table 7.4 Results obtained with the function boot.sw98() from the FEAR package

	1	2	3	4	5
dhat	1.00	1.60	1.00	1.11	1.00
dhat.m	0.86	1.45	0.87	1.01	0.85
bias	−0.14	−0.15	−0.13	−0.10	−0.15
dhat.bc	1.14	1.75	1.13	1.21	1.15
ci.low	1.01	1.62	1.01	1.12	1.01
ci.high	1.32	1.94	1.28	1.36	1.38

7.7 Recommended Reading

The stochastic data envelopment approach has been developed by Simar and Wilson in a series of papers. A comprehensive overview is provided by the authors in

- Simar L, Wilson PW (2000) Statistical inference in nonparametric frontier models: The state of the art. J Prod Anal 13:49–78.

A concise description of the algorithm which is implemented in the FEAR package is given in

- Simar L, Wilson PW (1998) Sensitivity analysis of efficiency scores: How to bootstrap in nonparametric frontier models. Manag Sci 44(1):49–61.

Detailed information on the FEAR package is provided in

- Wilson PW (2008) Fear 1.0: A software package for frontier efficiency analysis with r. Soc Econ Plann Sci 42:247–254.

A comprehensive overview on the topic and recent developments is given in

- Simar P, Wilson PW (2007) Statistical inference in nonparametric frontier models: Recent developments and perspectives. In: Fried HO, Lovell CAK, Schmidt SS (eds) The measurement of productive efficiency and productivity growth. Oxford University Press, Oxford, pp 421–521.

7.8 Exercises

Consider the following input and output data for 8 DMUs (Table 7.5):

1. Display the inputs and outputs graphically.
2. Estimate efficiency scores (potential output increase given input quantity) under the assumption of variable returns to scale.
3. Add the efficient frontier to your plot.
4. Indicate the potential output increase for inefficient firms with arrows.
5. Assume that a sample of size $n = 4$ shall be drawn. How many different samples are possible?
6. Consider the specific sample s with firm indices $\{2, 3, 6, 7\}$. Obtain the efficiency scores for the four firms and calculate the bias.
7. Assume that the 8 firms given in the example are a random sample drawn from all $N = 20$ firms operating in a specific sector. Would you expect your efficiency

Table 7.5 Example

	A	B	C	D	E	F	G	H
x: Input	1	2	3	4	5	6	7	8
y: Output	1	2	2	7	5	6	4	8

scores to be biased? How would you assess the situation if there were only 8 firms in the sector and you observed the complete population?

8. Draw $B = 100$ samples of size $n = 4$ without replacement and calculate the efficiency scores for each sample. Compare the average efficiency scores obtained from samples of size $n = 4$ with your original efficiency scores.
9. Describe the strongly simplified version (see Sect. 7.6.1) of the approach suggested by Simar and Wilson.
10. Apply the strongly simplified version regarding the 8 firms as a sample from a superpopulation. Note that you have to consider the reciprocal of the input-oriented efficiency scores instead of output-oriented scores to make the algorithm work.

References

Simar L, Wilson PW (1998) Sensitivity analysis of efficiency scores: How to bootstrap in nonparametric frontier models. Manag Sci 44(1):49–61

Simar L, Wilson PW (2000) Statistical inference in nonparametric frontier models: The state of the art. J Prod Anal 13:49–78

Simar P, Wilson PW (2007) Statistical inference in nonparametric frontier models: Recent developments and perspectives. In: Fried HO, Lovell CAK, Schmidt SS (eds) The measurement of productive efficiency and productivity growth. Oxford University Press, Oxford, pp 421–521

Wilson PW (2008) Fear 1.0: A software package for frontier efficiency analysis with r. Soc Econ Plann Sci 42:247–254

Stochastic Frontier Analysis

8

Abstract

The stochastic frontier analysis is an econometric approach to efficiency measurement. The basic idea is the introduction of two error components, a random error term and an inefficiency term. For both terms, a distributional assumption is made, which facilitates maximum likelihood estimation.

8.1 Introduction

Stochastic frontier analysis has been introduced in the literature by Aigner et al. (1977) and Meeusen and van de Broeck (1977). The basic idea of the stochastic frontier analysis is the introduction of an composed error term which is the sum of a purely random (usually normal) error term and a stochastic inefficiency term which is assumed to be generated with a specific random distribution (e.g., the exponential or the half-normal).

8.2 Production and Inefficiency

We introduce a deterministic production process, e.g., a production function we already discussed, which is overlaid by two different stochastic terms.

A. Behr, *Production and Efficiency Analysis with R*,
DOI 10.1007/978-3-319-20502-1_8

8.2.1 Observed and Efficient Production

The basic idea of efficiency analysis is the comparison of an observed output y and an output $\tilde{y}$ that would result if inputs were used "efficiently". The ratio

$$\eta = \frac{\text{observed production}}{\text{efficient production}} = \frac{y}{\tilde{y}} \tag{8.1}$$

serves as a measure of efficiency. The reciprocal of η indicates the gain in output when moving from inefficient to efficient production.

Efficiency literature is focused mainly on the question of how to obtain an estimate of the unknown $\tilde{y}$. While there have been many different methods proposed, two methods dominate the literature: data envelopment analysis (DEA) and stochastic frontier analysis (SFA). Usually, these two approaches are characterized as nonparametric and deterministic (DEA) on one hand and stochastic and parametric (SFA) on the other hand. In this chapter we discuss the SFA.

8.2.2 Production Frontier and Deviations

We use simple models and random number generators to fix ideas. While this might help thinking about production and efficiency, at this stage of reasoning, we do not refer to any real production process taking place in firms. We first propose a parametric relation between inputs x and efficient output $\tilde{y}$ by means of a production function

$$\tilde{y} = f(x, \beta). \tag{8.2}$$

To keep the model simple, we assume a very specific one-factor production function of the following form

$$\tilde{y} = x^{\beta}. \tag{8.3}$$

A firm i might produce output y_i that deviates from the efficient output $\tilde{y}$ for a variety of reasons. We can think of a deviation u_i resulting from the fact that producer i uses his inputs in an inefficient way.[1] Additionally we can imagine a "random" deviation v_i that might result in higher or lower output compared to $\tilde{y}$ because of influences producer i cannot control.[2] This results in the introduction of a combined error term in the production function $e^{\epsilon} = e^{-u} e^{v} = e^{v-u}$.

[1]For example a winemaker cutting his grapevines too rigorously and thereby lowering his harvest.

[2]For example, the weather conditions could be either advantageous or disadvantageous for the winemaker.

Overlaying a simple one-factor deterministic production function with the combined error term results in

$$y = x^{\beta}\, \mathrm{e}^{-u}\, \mathrm{e}^{v}, \quad \text{with } u > 0. \tag{8.4}$$

Taking logs, we obtain

$$\ln y = \beta \ln x - u + v. \tag{8.5}$$

8.2.3 Data Generation

We generate some artificial data according to the normal-half normal model using a simple one-factor production function:

$$\ln y = \beta_0 + \beta_1 \ln x + \epsilon = \beta_0 + \beta_1 \ln x + v - u. \tag{8.6}$$

Observations will be generated using the following setting

$$\begin{aligned} &\ln x = \{0.05, 0.1, \dots, 1\}, \quad \beta_0 = 0.2, \quad \beta_1 = 0.5, \\ &v \overset{\text{i.i.d.}}{\sim} \mathcal{N}(0, 0.01), \quad u \overset{\text{i.i.d.}}{\sim} \mathcal{N}^{+}(0, 0.04), \quad n = 20. \end{aligned} \tag{8.7}$$

With normally distributed error terms v and positive normally distributed inefficiency u terms the marginal density of the combined error term $\epsilon = v - u$ is

$$f(\epsilon) = \frac{2}{\sigma}\phi\left(\epsilon\sigma^{-1}\right)\left[1 - \Phi\left(\epsilon\lambda\sigma^{-1}\right)\right], \text{ where } \sigma^2 = \sigma_v^2 + \sigma_u^2 \text{ and } \lambda = \frac{\sigma_u}{\sigma_v}. \tag{8.8}$$

The density distribution of ϵ is asymmetric and characterized by

$$\mathrm{E}(\epsilon) = \mathrm{E}(v - u) = \mathrm{E}(v) - \mathrm{E}(u) = -\mathrm{E}(u) = -\frac{\sqrt{2}}{\sqrt{\pi}}\sigma_u. \tag{8.9}$$

and

$$\mathrm{Var}(\epsilon) = \sigma_\epsilon^2 = \mathrm{Var}(u) + \mathrm{Var}(v) = \left(\frac{\pi - 2}{\pi}\right)\sigma_u^2 + \sigma_v^2. \tag{8.10}$$

The generation of the output values y proceeds in the following steps:

1. Obtain the deterministic logarithmic output $\beta_0 + \beta_1 \ln x$ displaying the true production relation in the absence of both inefficiency and random noise.
2. Generate random inefficiency terms and random noise according to the distributional assumptions.

3. Add the half-normal inefficiency terms $-u$ and the random noise v resulting in the $\ln y$-values.

We first generate the error terms and display the density distributions of random noise v, the inefficiency term u, and the combined error term $\epsilon = v - u$.

```
s_u <- 0.2
s_v <- 0.1
s_2 <- s_u^2 + s_v^2
s <- sqrt(s_2)
l <- s_u/s_v
s_e_2 <- (pi-2)/pi*s_u^2 + s_v^2
s_e <- sqrt(s_e_2)
xx <- (seq(-2,2, by=0.001))
# function for density of v-u
fe <- function(x,s_u,s_v){
      s <- sqrt(s_u^2 + s_v^2)
      l <- s_u/s_v
      2/s*dnorm(x/s)*pnorm(-x*l/s)
      }
# realizations of u and v
set.seed(3)
n <- 20
u <- abs(rnorm(n,sd=s_u))
expu <- exp(-u)
v <- rnorm(n,sd=s_v)
expv <- exp(v)
```

In Fig. 8.1 we display the three distributions and indicate the realizations by small ticks.

```
# noise
plot(xx,dnorm(xx,0,sd=s_v),type="l",xlim=c(-1,1),
      lwd=1, col=1,xlab="v",ylab="f(v)")
rug(v)
# inefficiency
plot(xx,2*dnorm(xx,0,sd=s_u)*(xx>=0),type="l",xlim=c(-0.2,2),
      lwd=1,col=1,xlab="u",ylab="f(u)")
rug(u)
# combined error
plot(xx,fe(xx,s_u,s_v),type="l",xlim=c(-2,2),
      lwd=1,col=1,xlab="v-u",ylab="f(e)")
rug(v-u)
```

The input $\ln x$ and outputs $\ln y$ are generated as follows:

```
lx <- seq(1/n,1,length=n)
x <- exp(lx)
b0 <- 0.2
b1 <- 0.5
lxb <- b0+b1*lx
xb <- exp(lxb)
```

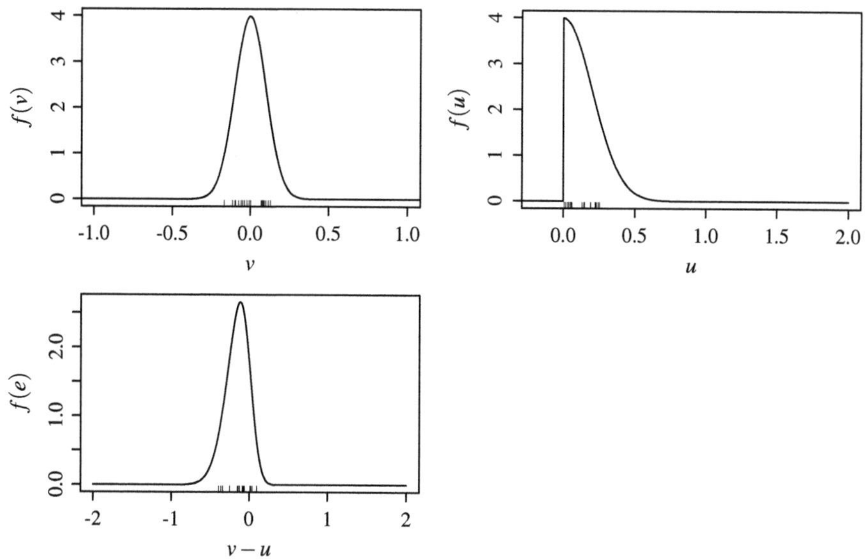

Fig. 8.1 Artificial production data: SFA, normal-half normal. *Top*: distributions of error term v (*left*) and inefficiency term u (*right*), *bottom*: scatter plot of logarithmic values with true frontier (combined error $(v-u)$)

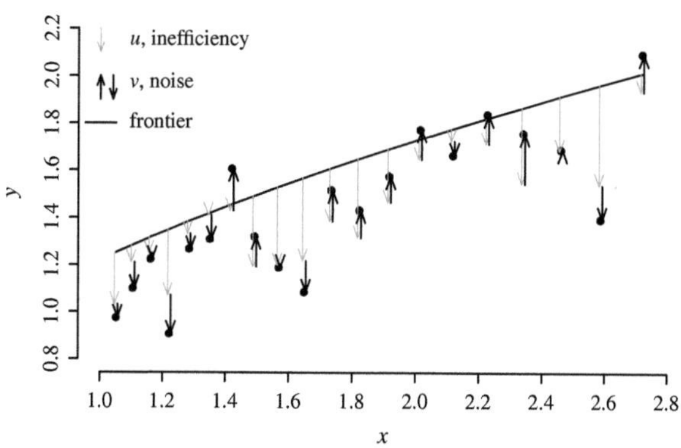

Fig. 8.2 Artificial production data: The production frontier and output including two error components

```
ly <- lxb-u+v
y <- exp(ly)
xb.u <- xb*expu
lxb.u <- lxb-u
```

We next display the model graphically using generated input and output in Fig. 8.2.

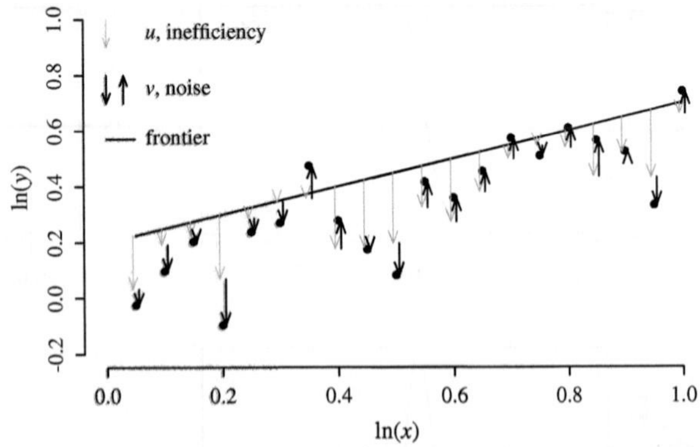

Fig. 8.3 Logarithmic artificial production data

```
# data
plot(x,y,ylim=c(0.8,2.2),xlim=c(1,2.8),
     ylab="y",axes=F,pch=19,cex=0.8)
axis(1,at=seq(1,2.8,0.2))
axis(2,at=seq(0.8,2.2,0.2))
# model and deviations
lines(x,xb)
arrows(x-0.005, xb, x-0.005, xb.u, length=0.1,
       angle=20,col="grey70")
arrows(x+0.005, xb.u, x+0.005, y, length=0.1, angle=20, col=1)
# legend
arrows(1,2.2,1,2.1, length=0.08, angle=20,col="grey70")
arrows(1,1.9,1,2, length=0.08, angle=20, col=1)
arrows(1.04,2,1.04,1.9, length = 0.08, angle = 20, col=1)
segments(0.95,1.8,1.05,1.8)
text(1.06,2.15,"u, inefficiency", pos=4)
text(1.06,1.97,"v, noise", pos=4)
text(1.06,1.8,"frontier", pos=4)
```

The model in logarithms is depicted in Fig. 8.3.

```
# data
plot(lx,ly,ylim=c(-0.2,1),xlim=c(0,1),pch=19,
     ylab="ln(y)",xlab="ln(x)",axes=F,cex=0.8)
axis(1,at=seq(0,1,0.2))
axis(2)
# model in logs and deviations
lines(lx,lxb)
arrows(lx-0.005, lxb, lx-0.005, log(xb.u),
       length=0.1, angle=20, col="grey70")
arrows(lx+0.005, log(xb.u), lx+0.005, ly,
       length=0.1, angle=20, col=1)
# legend
arrows(0,1,0,0.9, length=0.08, angle=20, col="grey70")
arrows(0,0.8,0,0.7, length=0.08, angle=20, col=1)
arrows(0.03,0.7,0.03,0.8, length=0.08, angle=20, col=1)
segments(0,0.57,0.05,0.57)
```

```
text(0.05,0.95,"u, inefficiency", pos=4)
text(0.05,0.75,"v, noise", pos=4)
text(0.05,0.57,"frontier", pos=4)
```

8.2.4 Corrected Ordinary Least Squares

When estimating the production function with empirical data, we usually assume that the deterministic relation between output and inputs $f(x_i, \beta)$ is overlaid by an error term. In ordinary least squares (OLS) regression analysis, we are used to estimate an average relationship between x and y by minimizing the sum of squared residuals. In this case, the estimated quantity of production $\hat{y}$ given the observed quantity of input x allows to calculate the difference between observed and estimated input $y_i - \hat{y}_i$. In the context of efficiency measurement, we have to keep in mind that this means that we use the conditional "average" output given the observed amount of input as a yardstick to judge the observed output. Because the estimated conditional "average" output does not comply with our idea of efficient output given the observed input, several approaches to estimate the production frontier have been discussed. Efficient production $\tilde{y}$ implies maximal possible production given the quantity of inputs. In the deterministic approach, the deviation between observed output y and output at the frontier $\tilde{y}$ is regarded as inefficiency. In this approach, we neglect the possibility of random errors and regard deviations from the frontier as coming from inefficient production only. A plausible idea in this case is to shift the estimated OLS-regression outwards by the maximal deviation between observed output y and fitted regression values $\hat{y}$

$$\tilde{y}^{\text{cols}} = \hat{y} + \max(y - \hat{y}). \tag{8.11}$$

Efficiency scores $\hat{\eta}$ are then calculated as

$$\hat{\eta} = \frac{y}{\tilde{y}^{\text{cols}}}. \tag{8.12}$$

The estimated frontier, obtained by shifting the OLS-regression, is also called corrected OLS (COLS) regression. In Fig. 8.3, we show the OLS as well as the shifted OLS regression line. In our model setup including an inefficient term u and an error term v, COLS will lead to somewhat misleading efficiency scores because negative random errors will cause an overestimation of inefficiency and positive random errors accordingly an underestimation.

8.2.5 Corrected Ordinary Least Squares Using R

The corrected ordinary least squares approach (COLS) is obviously not appropriate to the generation of data according to the SFA approach. Despite the fact that the

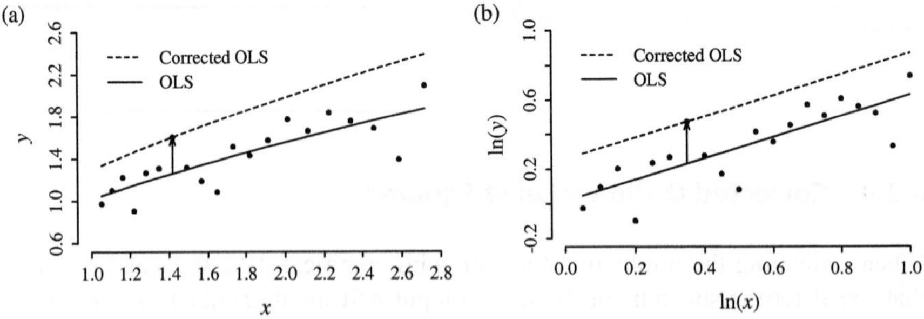

Fig. 8.4 Artificial production data: OLS and shifted OLS-regression line. (**a**) Original. (**b**) In logs

COLS approach ignores the existence of pure random noise, we run through a small example. We first estimate a simple OLS-regression, then we pick the highest residual and finally we shift the OLS-regression outwards. Because of ignoring the random noise, the COLS-approach results in an overestimation of inefficiency. This can be seen in Fig. 8.4.

We obtain the shifted regression by the following code:

```
reg <- lm(ly~lx)
sreg <- summary(reg)
yd <- exp(fitted(reg))
lyd <- fitted(reg)
yd.cols <- yd * max(y/yd)
lyd.cols <- lyd + max(ly-lyd)
no <- which(y/yd==max(y/yd))
```

Figure 8.4, which is obtained by the following code, displays the results graphically:

```
# original data
plot(x,y,ylim=c(0.6,2.6),xlim=c(1,2.8),
     ylab="y",pch=19,axes=F,cex=0.8)
axis(2,at=seq(0.6,2.6,0.4))
axis(1,at=seq(1,2.8,0.2))
# OLS/COLS
lines(x,yd)
lines(x,yd.cols,lty=2)
# legend
legend(1,2.6,c("Corrected OLS","OLS"),
       col=c(1,1),lty=c(2,1),bty="n")
arrows(x[no], yd[no], x[no], yd.cols[no],
       length = 0.1, angle = 20)

# data in logs
plot(lx,ly,ylim=c(-0.2,1),xlim=c(0,1),pch=19,
     ylab="ln(y)",xlab="ln(x)",axes=F,cex=0.8)
axis(1,at=seq(0,1,0.2))
axis(2)
```

```
# OLS/COLS
lines(lx,lyd)
lines(lx,lyd.cols,lty=2)
# legend
legend(0,1,c("Corrected OLS","OLS"),col=c(1,1),
       lty=c(2,1),bty="n")
arrows(lx[no], lyd[no], lx[no], lyd.cols[no],
       length = 0.1, angle = 20)
```

8.3 Maximum Likelihood Estimation

The stochastic frontier analysis is based on two error components (u and v) which have been already introduced above. To allow a maximum likelihood estimation of the production frontier, one has to make strong distributional assumptions. Typical assumptions about the distribution of the error components are that the noise term v follows the normal distribution and that inefficiency term $u > 0$ follows the half-normal distribution. Sometimes other assumptions for the inefficiency term are made, e.g., truncated normal, exponential, gamma, $\ldots$. In the following we stick to the normal-half normal assumption

$$u \overset{\text{i.i.d.}}{\sim} \mathcal{N}^+(0, \sigma_u^2) \quad \text{and} \quad v \overset{\text{i.i.d.}}{\sim} \mathcal{N}(0, \sigma_v^2). \tag{8.13}$$

We can think of the data generating process as a three step procedure:

1. The deterministic efficient output according to the production model is generated.
2. The efficient output is reduced because of inefficiency $-u$.
3. The inefficient output is overlaid by noise v.

The inefficiency term u is positive by assumption. Therefore, the composed error term $\epsilon = v - u$ has a skewed distribution and an expected value less than 0. Because of $\mathrm{E}(\epsilon = v - u) \neq 0$, OLS estimation results in biased estimates.

In the simple Cobb–Douglas case, we obtain the following function

$$y_i = \mathrm{e}^{\epsilon_i} \prod_{k=1}^{K} x_{ki}^{\beta_k}, \quad i = 1, \ldots, n. \tag{8.14}$$

In logs the production function is

$$\ln y_i = \sum_{k=1}^{K} \beta_k \ln x_{ki} + \epsilon_i = \ln(x_i')\beta + \epsilon_i. \tag{8.15}$$

8.3.1 The Log-Likelihood

Because of the additive error term, we have to deal with the two densities: The density of the noise $f_v(y)$ and the density of the inefficiency $f_u(x)$. Assuming independence, results in $f_{u,v}(x,y) = f_u(x)f_v(y)$. Note that the joint density of ϵ and u can be written as the product of a conditional and an unconditional density

$$f_{u,\epsilon}(x,z) = f_{\epsilon|u}(z|x) \cdot f_u(x). \tag{8.16}$$

The marginal density $f_\epsilon(z)$ of $\epsilon = v - u$ can be obtained from the joint density $f_{u,\epsilon}(x,z)$ through integrating out x

$$f_\epsilon(z) = \int_0^\infty f_{\epsilon|u}(z|x) \cdot f_u(x)\,\mathrm{d}x. \tag{8.17}$$

For the normal-half normal case, we obtain the marginal density as

$$\begin{aligned} f(\epsilon) &= \frac{2}{\sigma}\phi\left(\epsilon\sigma^{-1}\right)\left[1 - \Phi\left(\epsilon\lambda\sigma^{-1}\right)\right], \\ &\text{where } \sigma^2 = \sigma_v^2 + \sigma_u^2 \text{ and } \lambda = \frac{\sigma_u}{\sigma_v}. \end{aligned} \tag{8.18}$$

Note that this integration is somewhat tedious. The density distribution of ϵ is asymmetric and characterized by

$$\mathrm{E}(\epsilon) = \mathrm{E}(v-u) = \mathrm{E}(v) - \mathrm{E}(u) = -\mathrm{E}(u) = -\frac{\sqrt{2}}{\sqrt{\pi}}\sigma_u. \tag{8.19}$$

The variance of ϵ is given by

$$\mathrm{Var}(\epsilon) = \sigma_\epsilon^2 = \mathrm{Var}(u) + \mathrm{Var}(v) = \left(\frac{\pi-2}{\pi}\right)\sigma_u^2 + \sigma_v^2. \tag{8.20}$$

Based on the distribution of the composed error term $f(\epsilon)$ and the assumption of independence across firms, the log-likelihood is given by

$$\begin{aligned} \ln L(y|\beta,\lambda,\sigma^2,x) &= \ln \mathrm{P}\left(y_1 \cap \ldots \cap y_i \cap \ldots \cap y_n\right) \\ &= \ln \prod_{i=1}^n f(\epsilon_i) = \sum_{i=1}^n \ln f(\epsilon_i) \\ &= \sum_{i=1}^n \ln\left(\frac{2}{\sigma}\phi\left(\epsilon_i\sigma^{-1}\right)\left[1 - \Phi\left(\epsilon_i\lambda\sigma^{-1}\right)\right]\right). \end{aligned} \tag{8.21}$$

Using the expression for the standard normal distribution ϕ, we find

$$\begin{aligned}
\ln L(y|\beta,\lambda,\sigma^2,x) &= \sum_{i=1}^{n} \ln\left(\frac{2}{\sigma}\left(\frac{1}{\sqrt{2\pi}}\mathrm{e}^{-\frac{1}{2}\frac{\epsilon_i^2}{\sigma^2}}\right)\left[1-\Phi\left(\epsilon\lambda\sigma^{-1}\right)\right]\right) \\
&= \sum_{i=1}^{n} \ln\left(\frac{\sqrt{2}}{\sqrt{\pi}}\frac{1}{\sigma}\mathrm{e}^{-\frac{1}{2}\frac{\epsilon_i^2}{\sigma^2}}\left[1-\Phi\left(\epsilon\lambda\sigma^{-1}\right)\right]\right) \\
&= n\ln\left(\frac{\sqrt{2}}{\sqrt{\pi}}\right) - n\ln(\sigma) - \frac{1}{2\sigma^2}\sum_{i=1}^{n}\epsilon_i^2 \\
&\quad + \sum_{i=1}^{n}\ln\left[1-\Phi\left(\epsilon_i\lambda\sigma^{-1}\right)\right]
\end{aligned} \tag{8.22}$$

where the additive error term ϵ_i is replaced by $\ln y_i - \ln(x_i')\beta$ for estimation. Maximizing the log-likelihood results in the estimates $\hat{\beta}$, $\hat{\sigma}^2$, and $\hat{\lambda}$. Given these estimates

$$\hat{\sigma}^2 = \hat{\sigma}_u^2 + \hat{\sigma}_v^2 \text{ and } \hat{\lambda} = \frac{\hat{\sigma}_u}{\hat{\sigma}_v} \tag{8.23}$$

estimates of the variance components σ_u^2 and σ_v^2 can be obtained as follows

$$\hat{\sigma}_v^2 = \hat{\sigma}^2 - \hat{\sigma}_u^2 = \hat{\sigma}^2 - \hat{\lambda}^2\hat{\sigma}_v^2 = \frac{\hat{\sigma}^2}{(1+\hat{\lambda}^2)} \tag{8.24}$$

and

$$\hat{\sigma}_u^2 = \hat{\sigma}^2 - \frac{\hat{\sigma}^2}{(1+\hat{\lambda}^2)} = \frac{\hat{\sigma}^2(1+\hat{\lambda}^2)-\hat{\sigma}^2}{(1+\hat{\lambda}^2)} = \hat{\sigma}^2\frac{\hat{\lambda}^2}{(1+\hat{\lambda}^2)}. \tag{8.25}$$

8.3.2 Estimation of Individual Inefficiency Terms

Given the ML-parameter estimates, we can calculate estimated composed errors $\hat{\epsilon}$. Given the single estimate $\hat{\epsilon}_i$ for an individual firm i, it is impossible to obtain estimates for u and v for each individual firm.

But as the main objective of the analysis is the estimation of individual inefficiency scores η_i, we can calculate the conditional expectation of e^{-u} given the composed error term ϵ

$$\eta^* = \mathrm{E}\left(\mathrm{e}^{-u}\,|\epsilon\right) \tag{8.26}$$

where

$$\epsilon = \ln y - \ln(x')\beta. \tag{8.27}$$

The intuition is as follows: We want to estimate two quantities, inefficiency u and noise v, but obtain only an estimate of $\epsilon = v - u$. Obviously, we will not be able to obtain consistent estimates of u and v from a cross section (twice as many observations, twice as many u and v to estimate!). But we know $\mathrm{E}(v) = 0$ and $\mathrm{E}(-u) < 0$, therefore, if we obtain an $\hat{\epsilon} > 0$, we will rather assume that u is small (rather efficient firm) and if we obtain $\hat{\epsilon} << 0$, we will rather assume that u is big (rather inefficient firm).

For estimating the inefficiency term, we derive the expected value of the inefficiency term given the composed error term

$$\mathrm{E}\left[\mathrm{e}^{-u} \,|\epsilon = z\right]. \tag{8.28}$$

Therefore, we need the conditional distribution of u given ϵ

$$f_{u|\epsilon}(x|z) = \frac{f_{u,\epsilon}(x,z)}{f_\epsilon(z)}. \tag{8.29}$$

Given this distribution, the expected value can be derived

$$\begin{aligned} \mathrm{E}\left[\mathrm{e}^{-u} \,|\epsilon = z\right] &= \int_0^\infty \mathrm{e}^{-x} f_{u|\epsilon}(x|z)\,\mathrm{d}x = \int_0^\infty \mathrm{e}^{-x} \frac{f_{u,\epsilon}(x,z)}{f_\epsilon(z)}\,\mathrm{d}x \\ &= \frac{1}{f_\epsilon(z)} \int_0^\infty \mathrm{e}^{-x} f_{u,\epsilon}(x,z)\,\mathrm{d}x. \end{aligned} \tag{8.30}$$

The conditional density of u given ϵ in the normal-half normal case is

$$\begin{aligned} &f(u|\epsilon) = \sigma^{*-1}\phi\left(\frac{u-\mu^*}{\sigma^*}\right)\left[1-\Phi\left(\frac{-\mu^*}{\sigma^*}\right)\right], \quad \text{with} \quad u > 0, \\ &\text{where} \quad \mu^* = -\frac{\epsilon\sigma_u^2}{\sigma^2} = -\epsilon\gamma, \quad \sigma^* = \sqrt{\frac{\sigma_u^2\sigma_v^2}{\sigma^2}} = \sqrt{\sigma^2\gamma(1-\gamma)} \\ &\text{and} \quad \gamma = \sigma^{-2}\sigma_u^2. \end{aligned} \tag{8.31}$$

The inefficiency term u given the composed error term ϵ follows a normal distribution

$$u|\epsilon \sim \mathcal{N}^+(\mu^*, \sigma^*). \tag{8.32}$$

Based on the conditional distribution $f(u|\epsilon)$, the expected value of e^{-u} given ϵ is

$$E(e^{-u}|\epsilon) = \frac{1 - \Phi(\sigma^* - \mu^*/\sigma^*)}{1 - \Phi(-\mu^*/\sigma^*)} e^{-\mu^* + \frac{1}{2}\sigma^{*2}}. \tag{8.33}$$

Because we have introduced a stochastic error term v, we define the technical efficiency scores as

$$\eta = \frac{\text{factual production}}{\text{stochastic efficient production}} = \frac{f(x'; \beta)\, e^{v}\, e^{-u}}{f(x'; \beta)\, e^{v}} = e^{-u}. \tag{8.34}$$

In the case of efficient production ($u = 0$), we find $\eta = e^{-u} = e^{0} = 1$. We estimate $E(e^{-u}|\epsilon)$ using the ML-estimates

$$\hat{\sigma}^* = \frac{\hat{\sigma}_u^2 \hat{\sigma}_v^2}{\hat{\sigma}_u^2 + \hat{\sigma}_v^2} \quad \text{and} \quad \hat{\mu}^* = -\frac{\hat{\epsilon}\hat{\sigma}_u^2}{\hat{\sigma}_u^2 + \hat{\sigma}_v^2} \tag{8.35}$$

as

$$\hat{\eta} = \frac{1 - \Phi(\hat{\sigma}^* - \hat{\mu}^*/\hat{\sigma}^*)}{1 - \Phi(-\hat{\mu}^*/\hat{\sigma}^*)} e^{-\hat{\mu}^* + \frac{1}{2}\hat{\sigma}^{*2}}. \tag{8.36}$$

Note that there are some critical drawbacks with the stochastic frontier analysis. The estimation rests on strong distributional assumptions, which can hardly be justified by economic reasoning. The precision of the estimated individual inefficiency scores depends strongly on the ratio of the variance of the inefficiency term and the variance of the error term. The estimation procedure is highly sensitive towards individual outliers. The estimates of individual inefficiency terms are not consistent, that is with increasing n the mass of the distribution does not concentrate on the true individual inefficiency.

8.3.3 Maximum Likelihood with R

We have to maximize the log-likelihood function numerically to obtain the maximum likelihood estimates of the parameters of the production function β_0 and β_1 and of the variances σ_u^2 and σ_v^2. Given these estimates, we can calculate deviations $\hat{\epsilon}$ from the estimated productions frontier. And given these estimates, we finally calculate our estimates of the expected values of the inefficiency terms $E[e^{-u}]$ which we use as estimates of individual inefficiency e^{-u}.

The following expression shows that we have to maximize the likelihood through variation of a parameter vector θ containing four elements: $\theta' = (\beta_0, \beta_1, \lambda, \sigma^2)$.

$$\ln L(y|\beta, \lambda, \sigma^2, x) = n \ln\left(\frac{\sqrt{2}}{\sqrt{\pi}}\right) - n\ln(\sigma) - \frac{1}{2\sigma^2}\sum_{i=1}^{n}\epsilon_i^2 \\ + \sum_{i=1}^{n}\ln\left[1 - \Phi\left(\epsilon_i\lambda\sigma^{-1}\right)\right], \tag{8.37}$$

where the additive error term ϵ_i is replaced by $\ln y_i - \ln(x_i')\beta$ for estimation. We write a function returning the value of the log-likelihood for a given vector of parameters. To maximize the log-likelihood, we use the function `optim()`. Because by default `optim()` minimizes, we scale the objective function by -1.

```
logl <- function(theta){
     b0 <- theta[1]
     b1 <- theta[2]
     g  <- theta[3]
     s2 <- theta[4]
     lxd <- b0+b1*lx
     e <- ly-lxd
     pn <- pnorm(-e*g/sqrt(s2))
           n*log(sqrt(2/pi))-n*log(sqrt(s2))+
           sum(log(pn))-1/(2*s2)*sum(e^2)
}
thetastart <- c(b0,b1,1,s_2)
o <- optim(thetastart,logl,control=list(fnscale=-1))
tab <- round(o$par,3)
names(tab) <- c("b0","b1","lambda","s^2");tab

##      b0      b1 lambda     s^2
##   0.163   0.603  2.760   0.036
```

Figure 8.5 displays the shape of the log-likelihood through varying a single parameter keeping the other three parameters fixed at their maximum likelihood values.

```
thetastart <- o$par
vec <- seq(0.001,5,0.001)
lvec <- length(vec)
erg <- matrix(NA,lvec,4)
for (i in 1:lvec) {
  t1i <- c(vec[i],thetastart[2:4])
  t2i <- c(thetastart[1],vec[i],thetastart[3:4])
  t3i <- c(thetastart[1:2],vec[i],thetastart[4])
  t4i <- c(thetastart[1:3],vec[i])
  erg[i,] <- c(logl(t1i),logl(t2i),logl(t3i),logl(t4i))}
plot(vec,erg[,1],type="l",xlim=c(0,0.5),ylim=c(-60,20),
     xlab=expression(beta[0]),ylab="Log-likelihood")
plot(vec,erg[,2],type="l",xlim=c(0,1.2),ylim=c(-100,20),
```

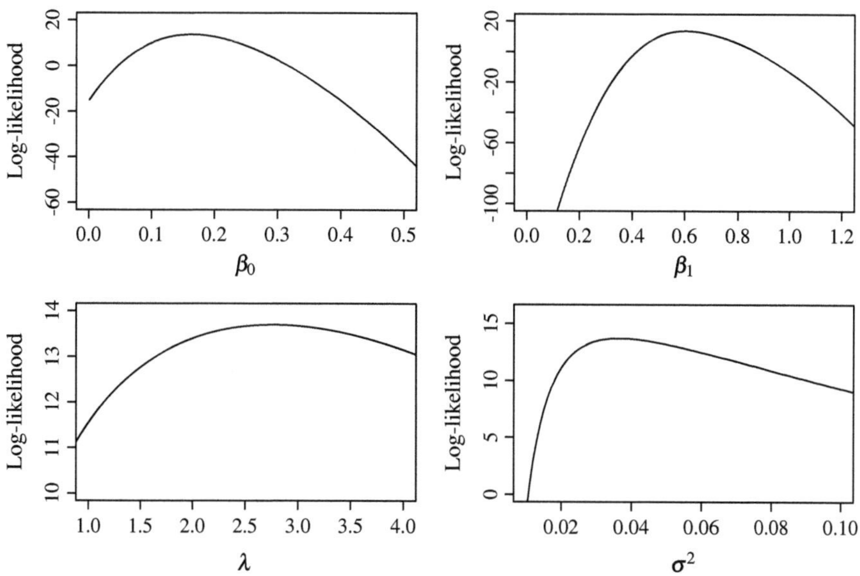

Fig. 8.5 Shape of the partial log-likelihoods

```
      xlab=expression(beta[1]),ylab="Log-likelihood")
plot(vec,erg[,3],type="l",xlim=c(1,4),ylim=c(10,14),
      xlab=expression(lambda),ylab="Log-likelihood")
plot(vec,erg[,4],type="l",xlim=c(0.01,0.1),ylim=c(0,16),
      xlab=expression(sigma^2),ylab="Log-likelihood")
```

To obtain the ML-estimates of the variances, we convert the estimated parameters σ^2 and λ to estimates of the variances σ_u^2 and σ_v^2.

```
b0.d <- o$par[1]
b1.d <- o$par[2]
g.d  <- o$par[3]
s2.d <- o$par[4]
s2.u.d <- s2.d*g.d^2/(1+g.d^2);s2.u.d

## [1] 0.03225

s2.v.d <- s2.d-s2.u.d;s2.v.d

## [1] 0.004234

tab <- round(c(b0.d,b1.d,g.d,s2.d,s2.u.d,s2.v.d),3)
names(tab) <- c("b0","b1","lambda","s^2","s_u^2","s_v^2");tab

##     b0     b1 lambda    s^2  s_u^2  s_v^2
##  0.163  0.603  2.760  0.036  0.032  0.004
```

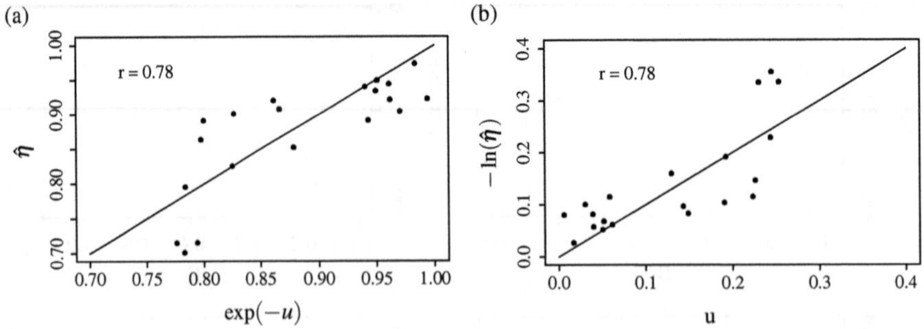

Fig. 8.6 Inefficiency scores. (**a**) True inefficiency scores $\exp(-u)$ and estimates $\hat{\eta}$. (**b**) u and $-\ln(\hat{\eta})$

Using the estimated regression parameters, we calculate the residuals of the regression and finally obtain the estimated efficiency scores $\hat{\eta}$:

```
ly.sfa <- b0.d+b1.d*lx
y.sfa <- exp(ly.sfa)
e.d <- ly-ly.sfa
mu.s <- -e.d*s2.u.d/s2.d
s.s <- sqrt(s2.u.d*s2.v.d/s2.d)
eta.d <- (1-pnorm(s.s-mu.s/s.s))/
         (1-pnorm(-mu.s/s.s))*exp(-mu.s+0.5*s.s^2)
```

Figure 8.6 shows the scatter plot of the true inefficiency scores $\exp(-u)$ and our estimates $\hat{\eta}$ (left panel) as well as the scatter plot for u and $-\ln(\hat{\eta})$ (right panel). Figure 8.6 is obtained by the following code.

```
plot(exp(-u),eta.d,ylab=expression(hat(eta)),
     xlim=c(0.7,1),ylim=c(0.7,1),pch=19,cex=0.8)
segments(0.7,0.7,1,1)
r <- cor(exp(-u),eta.d)
text(0.75,0.96,paste("r =",round(r,2)),cex=1.3)
plot(u,-log(eta.d),
     ylab=expression(paste("-ln(",hat(eta),")")),
     xlim=c(0,0.4),ylim=c(0,0.4),pch=19,cex=0.8)
segments(0,0,0.4,0.4)
r.u <- cor(u,-log(eta.d))
text(0.08,0.35,paste("r =",round(r.u,2)),cex=1.3)
```

A final comparison can be seen in Fig. 8.7 and is obtained by the following code. The regression has to be performed again, because we did not store the results in an object (same as in Sect. 8.2.5).

```
plot(lx,ly, xlim=c(0,1), ylim=c(0,1),
     ylab="ln(y)", xlab="ln(x)",axes=T,
     pch=19,cex=0.8)
lines(lx,lxb,col="darkgrey")
```

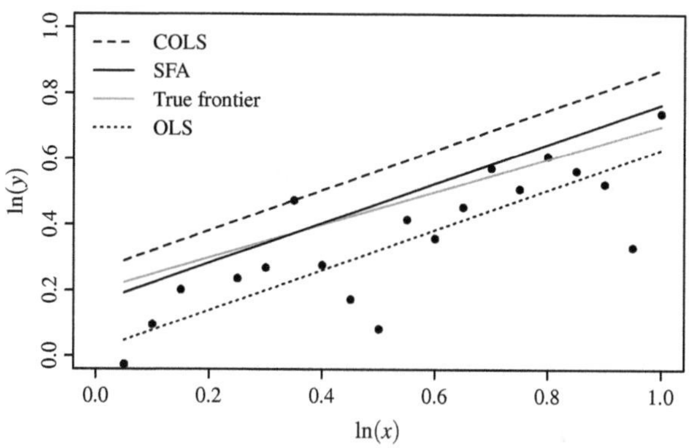

Fig. 8.7 Artificial production data: True frontier, SFA-frontier, OLS-regression, COLS-frontier

```
lines(lx,lyd.cols,lty=2)
lines(lx,fitted(reg),lty=3)
lines(lx,ly.sfa)
legend("topleft",c("COLS","SFA","True frontier","OLS"),
       lty=c(2,1,1,3),col=c(1,1,"darkgrey",1),bty="n")
```

8.4 Recommended Reading

The reference monograph on stochastic frontier analysis is

- Kumbhakar SC, Lovell CK (2000) Stochastic frontier analysis. Cambridge University Press, Cambridge.

The approach has been published almost simultaneously in two papers in 1977

- Aigner D, Lovell CK, Schmidt P (1977) Formulation and estimation of stochastic frontier production function models. J Econometrics 6:21–37.
- Meeusen W, van de Broeck J (1977) Efficiency estimation from Cobb–Douglas production functions with composed errors. Int Econ Rev 18:435–444.

A comprehensive overview on the topic and recent developments is provided by

- Greene WH (2008) The econometric approach to efficiency analysis. In: Fried HO, Lovell CAK, Schmidt SS (eds) The measurement of productive efficiency and productivity growth, chap 2. Oxford University Press, New York, pp 92–250.

8.5 Exercises

1. Consider the following stochastic production function

$$\ln y = \beta_0 + \beta_1 \ln x + \varepsilon = \beta_0 + \beta_1 \ln x + v - u$$

and generate a realization (`n=20, set.seed(123)`) using the following setup

$$\beta_0 = 0.3, \quad \beta_1 = 0.5, \quad \ln x \sim \mathscr{U}(0,1), \quad u \sim \mathscr{N}^+(0, 0.5^2),$$
$$v \sim \mathscr{N}(0, 0.3^2).$$

2. Estimate the conditional mean function to obtain $\hat{\beta}$ and $\hat{\sigma}^2$.
3. Estimate the deterministic frontier (corrected ordinary least squares) assuming $v = 0$ and using

$$y_i^{\text{cols}} = \hat{y}_i + \max\{y_i - \hat{y}_i | i = 1, \ldots, n\}.$$

4. Stochastic frontier analysis.
Consider the following likelihood

$$\ln L(y|\beta, \lambda, \sigma^2) = n \ln\left(\sqrt{\frac{2}{\pi}}\right) - n \ln(\sigma) - \frac{1}{2\sigma^2}\sum_{i=1}^{n} \varepsilon_i^2$$
$$+ \sum_{i=1}^{n} \ln\left[1 - \Phi\left(\varepsilon_i \lambda \sigma^{-1}\right)\right]$$

where

$$\sigma^2 = \sigma_v^2 + \sigma_u^2 \text{ and } \lambda = \frac{\sigma_u}{\sigma_v}$$

and the expected value of the inefficiency u given the combined error term ε

$$\mathrm{E}(\mathrm{e}^{-u} | \varepsilon) = \frac{1 - \Phi(\sigma^* - \mu^*/\sigma^*)}{1 - \Phi(-\mu^*/\sigma^*)} \mathrm{e}^{-\mu^* + \frac{1}{2}\sigma^{*2}}$$

$$\mu^* = -\frac{\varepsilon \sigma_u^2}{\sigma^2} = -\varepsilon\gamma; \quad \sigma^* = \sqrt{\frac{\sigma_u^2 \sigma_v^2}{\sigma^2}}.$$

a. Write a function returning a vector containing the start values for the parameter vector of the maximization problem $\theta = (\beta_0, \beta_1, \lambda, \sigma^2)'$ using the OLS-estimates for β and for σ^2 and the ad hoc chosen value $\lambda = 1$ based on observed $\ln y$ and $\ln x$.

b. Write a function returning the log-likelihood value for a given parameter vector θ (intern a).
c. Write a function returning the estimates of $\mathrm{E}(\mathrm{e}^{-u}\,|\varepsilon)$ for a given parameter vector θ (intern a).
d. Obtain the ML-estimator $\hat{\theta}_{\mathrm{ML}}$ using the function `optim()`.
e. Compare graphically the OLS regression and the stochastic production frontier (SFA).
f. Obtain estimates of the individual inefficiency e^{-u}.
g. Compare graphically the true inefficiencies e^{-u} and the estimated inefficiencies.

References

Aigner D, Lovell CK, Schmidt P (1977) Formulation and estimation of stochastic frontier production function models. J Econometrics 6:21–37

Greene WH (2008) The econometric approach to efficiency analysis. In: Fried HO, Lovell CAK, Schmidt SS (eds) The measurement of productive efficiency and productivity growth, chap 2. Oxford University Press, New York, pp 92–250

Kumbhakar SC, Lovell CK (2000) Stochastic frontier analysis. Cambridge University Press, Cambridge

Meeusen W, van de Broeck J (1977) Efficiency estimation from Cobb–Douglas production functions with composed errors. Int Econ Rev 18:435–444

Panel Data Stochastic Frontier Analysis

9

Abstract

Stochastic frontier analysis based on cross-sectional data is hampered by the fact that only one observation is available for the estimation of two error components. Panel data containing several observations for each firm considerably improve the situation for estimating firm specific efficiency scores if some assumptions on the time path of inefficiencies are introduced.

9.1 Introduction

The stochastic frontier analysis (SFA) is widely used to estimate firm individual efficiency scores and has been suggested by Aigner et al. (1977) and Meeusen and van de Broeck (1977). The basic idea is the introduction of an additive error term consisting of a noise and an inefficiency term. For the error as well as the inefficiency term, distributional assumptions are made, most often the normal and half-normal assumption, i.e., the error term (noise) is assumed to follow the normal distribution and the efficiency term is assumed to be half-normally distributed. Early discussions of using panel data sets for efficiency analysis have been provided by Pitt and Lee (1981) and Schmidt and Sickles (1984). Greene (2005) provides a detailed discussion of fixed and random effects in stochastic frontier models.

While typical panel data models assume a time invariant "inefficiency effect", this assumption must be regarded as highly unrealistic, as it excludes by assumption the possibility for firms to react on inefficiencies. Furthermore, if disregarded in the estimation procedure, firm specific time invariant heterogeneity may be "masquerading" as inefficiency. This problem is discussed in detail by Greene (2005).

A. Behr, *Production and Efficiency Analysis with R*,
DOI 10.1007/978-3-319-20502-1_9

9.2 Homogeneity and Firm Specific Time Invariant Inefficiency

In this section, we regard time invariant firm specific inefficiency u_i and assume that any firm specific effects α_i are absent. Furthermore, we use a very simple one input log-linear model of production. We index the N firms by i ($i = 1, \ldots, N$), each observed for T_i periods. To ease the exposition, we assume $T_i = T$ for all i. This assumption of a balanced data set could easily be dropped. To ease the exposition, in this section, we regard only one input and one output and we use capital letters for original values and small letters for logarithmic values. Hence, X_{it} denotes the amount of input used by firm i in period t to produce output Y_{it}.

The deterministic production frontier is given by function $e^{\beta_0} X_{it}^{\beta_1}$. Output Y_{it} is given by the deterministic production multiplied by a purely random term $e^{v_{it}}$ which is firm and time specific and multiplied by a time invariant firm specific inefficiency term e^{-u_i}:

$$Y_{it} = e^{\beta_0} X_{it}^{\beta_1} e^{v_{it}} e^{-u_i}. \tag{9.1}$$

Taking logarithms, we obtain

$$y_{it} = \beta_0 + \beta_1 x_{it} + v_{it} - u_i. \tag{9.2}$$

The inefficiency term e^{-u_i} therefore is

$$e^{-u_i} = \frac{Y_{it}}{e^{\beta_0} X_{it}^{\beta_1} e^{v_{it}}} \tag{9.3}$$

and its logarithm $-u_i$ is

$$-u_i = y_{it} - (\beta_0 + \beta_1 x_{it} + v_{it}). \tag{9.4}$$

Note that the model

$$y_{it} = \beta_0 + \beta_1 x_{it} + v_{it} - u_i \tag{9.5}$$

is formally identical to the panel data model without inefficiency terms u_i but firm specific effects α_i:

$$y_{it} = \beta_0 + \beta_1 x_{it} + v_{it} + \alpha_i. \tag{9.6}$$

Therefore, without taking distributional assumptions on v_{it} and u_i into account, a natural choice for estimating firm specific effects is the fixed effects panel data model.

For the generation of data according to the model, we make some distributional assumptions which we (for the moment) disregard in the estimation procedure.

The error term ε_{it} is assumed to be composed of two components

$$\varepsilon_{it} = v_{it} - u_i. \tag{9.7}$$

The error component v_{it} is assumed to be normally distributed

$$v_{it} \overset{\text{i.i.d.}}{\sim} \mathcal{N}(0, \sigma_v^2) \tag{9.8}$$

and the component u_i is assumed to be positive representing time invariant firm specific production inefficiency and assumed to follow the half-sided normal distribution:

$$u_i \overset{\text{i.i.d.}}{\sim} \mathcal{N}^+(0, \sigma_u^2). \tag{9.9}$$

Note that in R we use M instead of T:

```
N <- 9     # Number of firms
M <- 3     # Number of years
NM <- N*M     # Number of observations
i <- rep(1:N,each=M)     # firm identifier
m <- rep(1:M,N)          # year identifier
set.seed(123)
x <- sort(runif(NM))
us <- abs(rnorm(N,sd=0.2))
u <- rep(us,each=M)
v <- rnorm(NM,sd=0.05)
y <- 1+x+v-u
```

Figure 9.1 shows the distributions used to generate the inefficiencies ($-u$) and error terms (v) and the obtained realizations.

In Fig. 9.2, we show the true deterministic production function ($v = u = 0$) and the N linear production relations including the firm specific inefficiencies ($v = 0, u = u_i$). The observations are denoted by numbers indicating the number of the firm and contain inefficiencies (u_i) as well as noise (v_{it}).

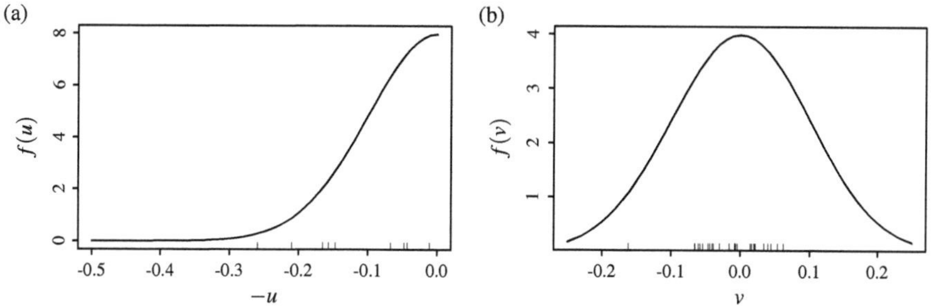

Fig. 9.1 Generating inefficiency and error terms. (**a**) Inefficiencies. (**b**) Error terms

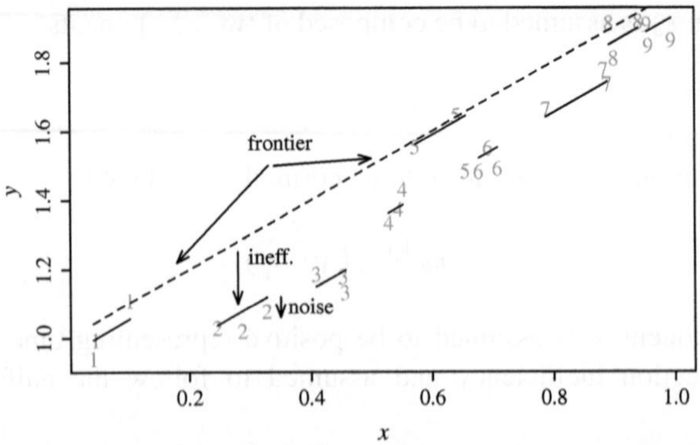

Fig. 9.2 Frontier (*dashed line*) and firm specific inefficient (*solid lines*) production functions

Note that with one individual intercept β_0 and N individual inefficiency terms, the model can be formulated as

$$\begin{aligned} y_{it} &= (\beta_0 - u_i) + \beta_1 x_{it} + v_{it} \\ &= \gamma_i + \beta_1 x_{it} + v_{it}, \quad \text{with} \quad \gamma_i = (\beta_0 - u_i). \end{aligned} \tag{9.10}$$

Because of complete linear dependence of an overall intercept and N dummy variables, in estimation we drop the first dummy variable (that is for firm $i = 1$) and use only $N - 1$ dummy variables in the regression with intercept:

$$\begin{aligned} y_{it} &= \beta_0^* + \gamma_2^* d_{2t} + \dots \gamma_j^* d_{jt} + \dots + \gamma_N^* d_{Nt} + \beta_1 x_{it} + v_{it} \\ d_{jt} &= \begin{cases} 1, & \text{if } j = i \\ 0, & \text{if } j \neq i. \end{cases} \end{aligned} \tag{9.11}$$

From the regression, we obtain the estimates $\hat{\beta}_0^*, \hat{\gamma}_2^*, \dots, \hat{\gamma}_N^*$. The estimates of the γ-parameters can be obtained as

$$\begin{aligned} \hat{\gamma}_1 &= \hat{\beta}_0^* \\ \hat{\gamma}_2 &= \hat{\beta}_0^* + \hat{\gamma}_2^* \\ &\vdots \\ \hat{\gamma}_N &= \hat{\beta}_0^* + \hat{\gamma}_N^*. \end{aligned} \tag{9.12}$$

```
# Dummy-regression
reg1 <- lm(y~x+as.factor(i))
g.s.h <- coef(reg1)
g.h <- c(g.s.h[1],g.s.h[-(1:2)]+g.s.h[1])
```

When using demeaned data $\tilde{x}_{it} = x_{it} - \bar{x}_i$ and $\tilde{y}_{it} = y_{it} - \bar{y}_i$, we obtain $\hat{\beta}_1$ from the regression and the estimates $\hat{\gamma}_i$ as

$$\hat{\gamma}_i = \bar{y}_i - \hat{\beta}_1 \bar{x}_i. \tag{9.13}$$

```
# Demeaned data
x.t <- unlist(tapply(x,i,function(z)z-mean(z)))
y.t <- unlist(tapply(y,i,function(z)z-mean(z)))
reg2 <- lm(y.t~x.t)
coef(reg2)

## (Intercept)          x.t
##   5.342e-18   6.778e-01

y.f <- fitted(reg2)
y.m <- unlist(tapply(y,i,mean))
x.m <- unlist(tapply(x,i,mean))
g.h <- y.m-x.m*coef(reg2)[2]
```

Instead of demeaning the data using the `tapply()` command before running the regression, we could also use a panel data command from the plm-package:

```
library(plm)
d <- pdata.frame(data.frame(i,m,x,y))
reg3 <- plm(y~x, data=d, model="within", effect="individual")
```

Unfortunately the estimates $\hat{\gamma}_i$ are not the estimated $\hat{u}_i$ inefficiencies because $\gamma_i = (\beta_0 - u_i)$. Obviously we need the estimate $\hat{\beta}_0$ to obtain estimated inefficiencies. The inefficiency term $-u_i$ is negative and its theoretically maximal value is 0. Therefore, it is plausible to use the maximum of the estimated $\hat{\gamma}_i$ parameters as an estimate for β_0

$$\hat{\beta}_0 = \max\{\hat{\gamma}_j\}, \quad j = 1, \ldots, N. \tag{9.14}$$

We finally obtain the estimated inefficiency terms $\hat{u}_i$ as

$$\hat{u}_i = \hat{\beta}_0 - \hat{\gamma}_i = \max\{\hat{\gamma}_j\} - \hat{\gamma}_i, \quad j = 1, \ldots, N. \tag{9.15}$$

```
b0.d <- max(g.h)
u.h <- b0.d-g.h
```

In Fig. 9.3, we show the N true inefficiencies $-u$ and the estimated inefficiencies $-\hat{u}$. To estimate the firm-specific inefficiency term $-u_i$, which in this model is assumed to be constant across time, we can make use of the T_i observations for firm i. Therefore, the more time periods are available, the smaller the variance of the estimator $-\hat{u}_i$. Using the same set up but choosing $T_i = T = 10$ instead of

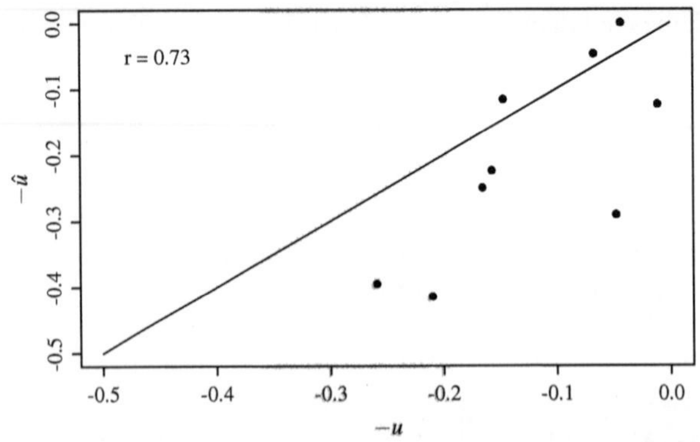

Fig. 9.3 Generated and estimated inefficiencies

$T_i = T = 3$, the correlation between true and estimated inefficiencies increases on average towards about 0.93. But of course the assumption of constant inefficiencies across time becomes the less plausible, the longer the time period T is.

9.3 Time Varying Firm Specific Inefficiency

So far, we assumed that the technical inefficiency is constant across time ($-u_{it} = -u_i$). As firms in a competitive environment will compare their own performance with its competitors, it is highly unlikely that inefficiency will be constant for several periods (e.g., years). Therefore, a natural extension of the model is to allow inefficiency to vary in time. This can be achieved by introducing a function of time $\eta_{it} = \beta_i(t)$:

$$e^{u_{it}} = e^{-u_i} e^{\beta_i(t)} = e^{-u_i+\beta_i(t)} = e^{-u_i+\eta_{it}}. \tag{9.16}$$

Starting from the model

$$Y_{it} = e^{\beta_0} X_{it}^{\beta_1} e^{v_{it}} e^{-u_i+\eta_{it}} \tag{9.17}$$

and taking logarithms, we obtain

$$y_{it} = \beta_0 + \beta_1 x_{it} + v_{it} - u_i + \eta_{it}. \tag{9.18}$$

The time specific inefficiency term $e^{-u_i+\beta_i(t)} = e^{-u_i+\eta_{it}}$, therefore, is

$$e^{-u_i+\eta_{it}} = \frac{Y_{it}}{e^{\beta_0} X_{it}^{\beta_1} e^{v_{it}}} \tag{9.19}$$

and its logarithm $-u_i + \eta_{it}$ is

$$-u_i + \eta_{it} = y_{it} - (\beta_0 + \beta_1 x_{it} + v_{it}). \tag{9.20}$$

As the spell length of time series observed for individuals is most often rather short, it is desirable to have a scarce parameterization when modeling the inefficiency as a function of time. Note that the model can be restricted to have an identical development of inefficiency in time, i.e., $\beta_i(t) = \beta(t)$. One specific scarce parameterization has been suggested by Battese and Coelli (1992) which introduces only one additional parameter for the time path[1]

$$e^{\eta_{it}} = e^{-\zeta_i(t-T)}, \quad t = 1, \ldots, T \tag{9.21}$$

which in logarithm simplifies towards

$$\eta_{it} = -\zeta_i(t-T), \quad t = 1, \ldots, T. \tag{9.22}$$

Note that in the final period observed ($t = T$), we have $\exp(\eta_{it}) = 1$ and $\eta_{it} = 0$. Therefore, $\exp(-u_i + \eta_{it}) = \exp(-u_i)$ and $-u_i + \eta_{it} = -u_i$. Despite its simplicity, the function $\exp(-\gamma_i(t-T))$ can generate a variety of time paths (see Fig. 9.4).

Inserting $\beta_i(t)$ into the logarithmic model, we obtain

$$y_{it} = \beta_0 + \beta_1 x_{it} + v_{it} - \gamma_i(t-T) - u_i. \tag{9.23}$$

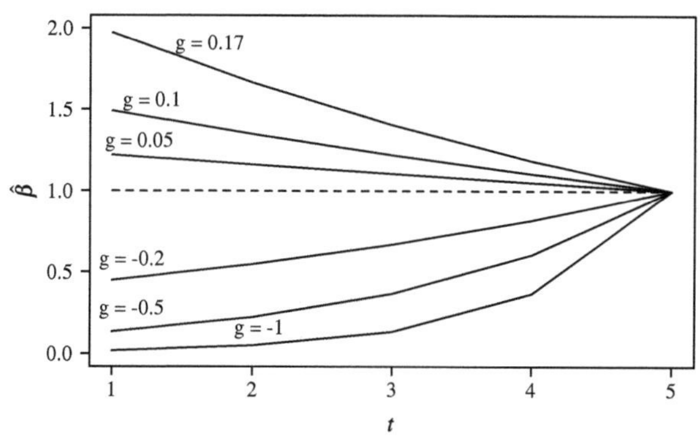

Fig. 9.4 Time path of inefficiency

[1] In the context of the stochastic frontier model relying on the assumption of normality, Battese and Coelli (1992) introduced a time varying inefficiency term of the form $\exp(-u_i\beta_i(t))$ convenient because of its simple effect on the parameters of the normal distribution. To obtain linearity in the parameters, we regard $\exp(-u_i + \beta_i(t))$ instead. Therefore, we assume a scaling of the exponentiated inefficiency term $\exp(-u)$ instead of $-u$.

The model can be estimated by OLS and differs only from the simple fixed effects model in the additional inclusion of a linear deterministic time trend.

We generate some artificial data according to the model and display the data graphically. To ease the interpretation, we use equidistant and sorted (increasing) values of the covariate x. Note that with non-sorted x-data, we would not obtain linear time trends in the plot (see Fig. 9.5).

```
N <- 6
NM <- N*M
i <- rep(1:N,each=M)
m <- rep(1:M,N)
tmT <- 1:M-M
tmTl <- rep(tmT,N)
set.seed(12)
x <- seq(0.1,0.99,length=NM)
us <- abs(rnorm(N,sd=0.2))
gv <- c(0.06,0.04,0.02,-0.02,-0.05,-0.1)
gvl <- rep(-gv,each=M)
b <- gvl*tmTl
u <- rep(us,each=M)
v <- rnorm(NM,sd=0.1)
y <- 1+x+v-u+b
```

With one individual intercept β_0 and N individual inefficiency terms, the model can be formulated as

$$\begin{aligned} y_{it} &= (\beta_0 - u_i) + \beta_1 x_{it} - \zeta_i(t-T) + v_{it} \\ &= \gamma_i + \beta_1 x_{it} - \zeta_i(t-T) + v_{it}, \qquad \text{where} \quad \gamma_i = (\beta_0 - u_i). \end{aligned} \tag{9.24}$$

Because of complete linear dependence of an overall intercept and N dummy variables, we drop the first dummy variable (that is for firm $i = 1$) in the regression

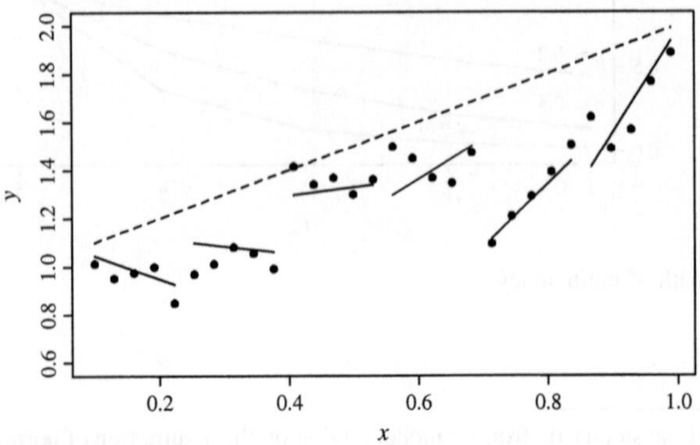

Fig. 9.5 Firm and time specific inefficiencies

with intercept.

$$\begin{aligned} y_{it} =& \beta_0^* + \beta_1 x_{it} + \gamma_2^* d_{2t} + \ldots \gamma_{j\cdot}^* d_{jt} + \ldots + \gamma_N^* d_{Nt} \\ & - \zeta_1^*(t-T) - \zeta_2^* d_{2t}(t-T) - \ldots - \zeta_N^* d_{Nt}(t-T) + v_{it} \\ d_{jt} =& \begin{cases} 1, & \text{if } j = i \\ 0, & \text{if } j \neq i. \end{cases} \end{aligned} \tag{9.25}$$

From the regression, we obtain the estimates $\hat{\beta}_0^*, \hat{\gamma}_2^*, \ldots, \hat{\gamma}_N^*, \hat{\zeta}_1^*, \ldots \hat{\zeta}_N^*$. The estimates of the firm and time period specific deviations α_{it} from the stochastic efficient frontier can be calculated for firm $i = 1$ as

$$\alpha_{it} = \hat{\beta}_0^* - \hat{\zeta}_1^*(t-T), \tag{9.26}$$

and for the $N-1$ remaining firms as

$$\alpha_{it} = \hat{\beta}_0^* + \gamma_i^* - (\hat{\zeta}_1 + \hat{\zeta}_i)(t-T). \tag{9.27}$$

The most efficient firm in period t is the firm with maximal value of α_{it} in period t. This firm is used as the benchmark for the $N-1$ other firms in this year. We finally obtain the estimated firm and period specific inefficiency terms $\hat{u}_{it}$ as

$$\hat{u}_{it} = \alpha_{it} - \max_i(\alpha_{it}|t = t'), \quad j = 1, \ldots, N, \quad t = 1, \ldots, T. \tag{9.28}$$

We estimate the time specific inefficiency terms using the linear regression function with dummy variables for firm effects and firm dummies interacted with the trend variable. Note that for the numerical example. We generate randomly input values x to prevent linear time dependence with the deterministic time trend as it was the case in the artificial example for the graphical exposition (see Fig. 9.6). For the numerical

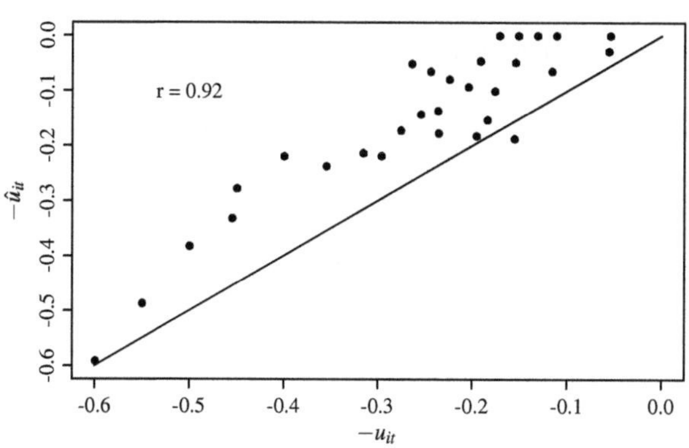

Fig. 9.6 Generated and estimated time varying inefficiency terms

example the correlation between generated and estimated firm specific time varying inefficiency terms is rather strong.

```
set.seed(123)
x <- runif(NM,0,1)
y <- 1+x+v-u+b
fi <- as.factor(i)
reg <- lm(y~x+fi+fi*tmT1)
ait <- fitted(reg)-coef(reg)[2]*x
ait.m <- tapply(ait,m,max)
u.h <- ait-rep(ait.m,N)
r <- round(cor(-u+b,u.h),2)
```

9.4 The Stochastic Frontier Model with Time Varying Inefficiency

So far, we estimated fixed effects panel data models and used the maximal firm specific fixed effect as the benchmark for the remaining firms. Both models with constant and time varying inefficiency have been estimated by ordinary least squares without any distributional assumptions of the noise and inefficiency terms. We now discuss a stochastic frontier model relying on the distributional assumptions of the noise and the efficiency term. The stochastic frontier analysis (SFA) is based on the pioneering work of Aigner et al. (1977) and Meeusen and van de Broeck (1977). Kumbhakar and Lovell (2000) provide a comprehensive overview. Because the SFA normal-half normal model can be regarded as the standard approach to efficiency estimation, we focus on this data generating process. The panel data model with a firm and time specific inefficiency term has been introduced by Battese and Coelli (1992).

A variation of the inefficiency term is allowed by introducing a function of time $\eta_{it} = \mathrm{e}^{-\eta(t-T)}$ which allows the inefficiency term to vary in time $U_{it} = \eta_{it} U_i = \mathrm{e}^{-\eta(t-T)}\, U_i$. Note that from now on we use uppercase letters for both random variates U and V and lowercase letters for their realizations. In original values the inefficiency term is exponentiated

$$\mathrm{e}^{-U_{it}} = \mathrm{e}^{-\eta_{it} U_i}\,. \tag{9.29}$$

Starting from a simple production model

$$Y_{it} = \mathrm{e}^{\beta_0}\, X_{it}^{\beta_1}\, \mathrm{e}^{V_{it}}\, \mathrm{e}^{-U_{it}}, \tag{9.30}$$

the time specific inefficiency term $e^{-U_{it}} = e^{-\eta_{it}U_i}$ is

$$e^{-U_{it}} = e^{-\eta_{it}U_i} = \frac{Y_{it}}{e^{\beta_0} X_{it}^{\beta_1} e^{V_{it}}}. \tag{9.31}$$

Taking logarithms, we obtain the model as

$$y_{it} = \beta_0 + \beta_1 x_{it} + V_{it} - U_{it} \tag{9.32}$$

and logarithm of the inefficiency term as

$$-U_{it} = -\eta_{it}U_i = y_{it} - (\beta_0 + \beta_1 x_{it} + V_{it}). \tag{9.33}$$

The specific scarce parameterization we discuss here has been suggested by Battese and Coelli (1992). The time path is determined by a single parameter

$$\eta_{it} = e^{-\eta_i(t-T)}, \quad t = 1, \ldots, T. \tag{9.34}$$

Note that in the final period observed ($t = T$), we have $\eta_{it} = e^0 = 1$ and, therefore, $\exp(-\eta_{it}U_i) = \exp(-U_i)$. The error term E_{it} is assumed to have two components:

$$E_{it} = V_{it} - U_{it}. \tag{9.35}$$

The error component V_{it} is assumed normal

$$V_{it} \overset{\text{i.i.d.}}{\sim} \mathcal{N}(0, \sigma_v^2) \tag{9.36}$$

and the component U_i representing time invariant production inefficiency is assumed to be (positive) truncated normal

$$U_i \overset{\text{i.i.d.}}{\sim} \mathcal{N}^+(\mu, \sigma_u^2). \tag{9.37}$$

9.4.1 Sketching the Idea of the Approach

Because the derivation of the log-likelihood is somewhat cumbersome, we first sketch the basic idea of deriving the log-likelihood. $f_U(u)$ and $f_V(v)$ denote the density functions of U and V, respectively. Because of the assumption of independence, the joint density function of U and V is $f_{U,V}(u, v) = f_U(u)f_V(v)$. Note that the joint density of E and U can be written as the product of a conditional and an unconditional density

$$f_{U,E}(u, e) = f_{E|U}(e|u) \cdot f_U(u). \tag{9.38}$$

Because of $(V-U)|_{U=u}$ being simply V shifted by a constant, the conditional density of $E|U$ is $f_V(v) = f_V(e+u) = f_{E|U}(e|u)$. The marginal density $f_E(e)$ of $E = V - U$ can be obtained from the joint density $f_{U,E}(u, e)$ through integrating out u

$$f_E(e) = \int_0^\infty f_{E|U}(e|u) \cdot f_U(u)\, du. \tag{9.39}$$

Since we only get an estimate of the combined error term by the maximum likelihood estimation, we have to derive an estimate for the inefficient term given the combined error term. Therefore, we derive the expected value of the inefficiency term given the composed error term

$$\mathrm{E}\left[e^{-u}|E = e\right]. \tag{9.40}$$

Therefore, we need the conditional distribution of U given E

$$f_{U|E}(u|e) = \frac{f_{U,E}(u,e)}{f_E(e)} \tag{9.41}$$

to derive the expected value $U|E$

$$\begin{aligned}\mathrm{E}\left[e^{-u}|E = e\right] &= \int_0^\infty \mathrm{e}^{-u} f_{U|E}(u|e)\, du = \int_0^\infty \mathrm{e}^{-u} \frac{f_{U,E}(u,e)}{f_E(e)}\, du \\ &= \frac{1}{f_E(e)} \int_0^\infty \mathrm{e}^{-u} f_{U,E}(u,e)\, du.\end{aligned} \tag{9.42}$$

9.4.2 Obtaining the Log-Likelihood

The density function of the positive truncated normal term U_i is

$$f_U(u) = \frac{\exp\left(-\frac{1}{2}\frac{(u-\mu)^2}{\sigma_u^2}\right)}{\sqrt{2\pi}\sigma_u\left[1 - \phi(-\mu/\sigma_u)\right]}, \quad u \geq 0. \tag{9.43}$$

The expected value of u_i and its variance are given by

$$\mathrm{E}[u] = \mu + \sigma_u \left\{\frac{\phi(-\mu/\sigma_u)}{1 - \Phi(-\mu/\sigma_u)}\right\} \tag{9.44a}$$

and

$$\mathrm{Var}[u] = \sigma_u^2 \left\{1 - \frac{\phi(-\mu/\sigma_u)}{1 - \Phi(-\mu/\sigma_u)}\left[\frac{\mu}{\sigma} + \frac{\phi(-\mu/\sigma_u)}{1 - \Phi(-\mu/\sigma_u)}\right]\right\}. \tag{9.44b}$$

We obtain the joint density as

$$f_{U,V}(u,v) = \frac{1}{\sqrt{2\pi}\sigma_u} \frac{\exp\left(-\frac{1}{2}\frac{(u-\mu)^2}{\sigma_u^2}\right)}{[1-\phi(-\mu/\sigma_u)]} \frac{1}{\sqrt{2\pi}\sigma_v} \exp\left(-\frac{1}{2}\frac{v^2}{\sigma_v^2}\right). \tag{9.45}$$

The joint density of $E_{it} = V_{it} - \eta_{it}U_i$ and U_i can be written as the product of a conditional and an unconditional density

$$f_{U,E}(u,e) = f_{E|U}(e|u) \cdot f_U(u). \tag{9.46}$$

Because of $(V-U)|_{U=u}$ being simply V shifted by a constant and the conditional density of $E|U$ being $f_V(v) = f_V(e+u) = f_{E|U}(e|u)$, we obtain the joint density as

$$\begin{aligned} f_{U_i,E_i}(u_i,e_i) &= f_{E_i|U_i}(e_i|u_i) \cdot f_{U_i}(u_i) \\ &= \frac{\exp\left\{\left(-\frac{1}{2}\frac{(u_i-\mu)^2}{\sigma_u^2}\right) + \frac{(e_i+\eta_i u_i)'(e_i+\eta_i u_i)}{\sigma_v^2}\right\}}{(2\pi)^{(T_i+1)/2}\sigma_u\sigma_V^{T_i}\,[1-\phi(-\mu/\sigma_u)]}. \end{aligned} \tag{9.47}$$

The marginal density $f_E(e)$ of $E = V - U$ can be obtained from the joint density $f_{U,E}(u,e)$ through integrating out u

$$\mathrm{E}[u] f_E(e) = \int_0^\infty f_{E|U}(e|u) \cdot f_U(u)\,\mathrm{d}u \tag{9.48a}$$

and for a specific i, we obtain

$$f_{E_i}(e_i) = \frac{\left[1-\phi\left(-\frac{\mu_i^*}{\sigma_{u_i}^*}\right)\right]\exp\left\{\frac{e_i'e_i}{\sigma_V^2} + \left(\frac{\mu}{\sigma_U}\right)^2 - \left(\frac{\mu_i^*}{\sigma_{U_i}^*}\right)^2\right\}}{(2\pi)^{(T_i)/2}\sigma_u\sigma_V^{(T_i-1)}\left[\sigma_V^2 + \eta_i'\eta_i\sigma_u^2\right]^{1/2}[1-\phi(-\mu/\sigma_u)]}, \tag{9.48b}$$

$$\text{where} \quad \mu_i^* = \frac{\mu\sigma_V^2 - \eta_i' e_i \sigma_U^2}{\sigma_V^2 + \eta_i'\eta_i\sigma_u^2} \quad \text{and} \quad \sigma_{U_i}^{*2} = \frac{\sigma_U^2\sigma_V^2}{\sigma_V^2 + \eta_i'\eta_i\sigma_u^2}. \tag{9.48c}$$

The density of U_i conditional on $E_i = e_i$ is given as

$$f_{U_i|E_i}(u_i|e_i) = \frac{1}{(2\pi)^{T_i/2}\sigma_{u_i}^*} \frac{\exp\left(-\frac{1}{2}\frac{(u_i-\mu_i^*)^2}{\sigma_u^{*2}}\right)}{\left[1-\phi(-\mu_i^*/\sigma_{U_i}^*)\right]}, \quad u_i \geq 0. \tag{9.49}$$

and therefore is the density of a positive truncated normal distribution with parameters μ_i^* and $\sigma_{U_i}^{*2}$. The expected value of the time varying error term $\mathrm{e}^{-\eta_{it}U_i}$ conditional on the combined error term is obtained through integration

$$
\begin{aligned}
\mathrm{E}\left[\mathrm{e}^{-\eta_{it}U_i}\,|E_i = e_i\right] &= \int_0^\infty \mathrm{e}^{-\eta_{it}u_i} f_{U_i|E_i=e_i}(u_i)\,\mathrm{d}u_i \\
&= \left(\frac{\left[1-\phi(\eta_{it}\sigma_{U_i}^* - \mu_i^*/\sigma_{U_i}^*)\right]}{\left[1-\phi(-\mu_i^*/\sigma_{U_i}^*)\right]}\right) \\
&\quad \times \exp\left(-\frac{1}{2}\frac{-\eta_{it}\mu_i^* + \frac{1}{2}\eta_{it}\sigma_{U_i}^{*2}}{\sigma_u^{*2}}\right).
\end{aligned} \tag{9.50}
$$

The expected value of $U_i|E_i = e_i$ is obtained as

$$
\begin{aligned}
\mathrm{E}\left[U_i|E_i = e_i\right] &= \int_0^\infty u_i f_{U_i|E_i=e_i}(u_i)\,\mathrm{d}u_i \\
&= \mu_i^* + \sigma_{U_i}^*\left\{\frac{\phi(-\mu_i^*/\sigma_{U_i}^*)}{1-\phi(-\mu_i^*/\sigma_{U_i}^*)}\right\}.
\end{aligned} \tag{9.51}
$$

The log-likelihood is given by

$$
\begin{aligned}
\ln L(\theta\,|y) = &\ \frac{1}{2}\left(\sum_{i=1}^N T_i\right)\ln(2\pi) - \frac{1}{2}\sum_{i=1}^N (T_i - 1)\ln(\sigma_V^2) \\
&- \frac{1}{2}\sum_{i=1}^N \ln(\sigma_V^2 + \eta_i'\eta_i\sigma_U^2) - N\ln\left[1-\Phi(-\mu/\sigma_u)\right] \\
&+ \sum_{i=1}^N \ln\left[1-\phi(-\mu_i^*/\sigma_{U_i}^*)\right] \\
&- \frac{1}{2\sigma_V^2}\sum_{i=1}^N (y_i - x_i\beta)'(y_i - x_i\beta) - \frac{1}{2}\left[N(\mu/\sigma_u)\right]^2 \\
&+ \frac{1}{2}\sum_{i=1}^N (\mu_i^*/\sigma_{U_i}^*)^2,
\end{aligned} \tag{9.52}
$$

with parameter vector $\theta = (\beta', \sigma_V^2, \sigma_U^2, \mu, \eta)'$. For programming, it is appropriate to use a slightly different parameterization

$$\sigma_S^2 = \sigma_V^2 + \sigma_U^2, \quad \gamma = \frac{\sigma_U^2}{\sigma_S^2} = \frac{\sigma_U^2}{\sigma_V^2 + \sigma_U^2}, \quad z = \frac{\mu}{\sqrt{\gamma\sigma_S^2}} = \frac{\mu}{\sigma_U}$$
$$\text{and} \quad z_i^* = \frac{\mu(1-\gamma) - \gamma\eta_i'(y_i - x_i\beta)}{\sqrt{\gamma(1-\gamma\sigma_S^2\left[\gamma(\eta_i'\eta_i - 1\right]}}. \tag{9.53}$$

Using this parameterization, we obtain the log-likelihood as

$$\begin{aligned}\ln L(\theta|y) = &\frac{1}{2}\left(\sum_{i=1}^{N} T_i\right)\{\ln(2\pi) + \ln(\sigma_S^2)\} - \frac{1}{2}\sum_{i=1}^{N}(T_i - 1)\ln(1-\gamma)\\ &- \frac{1}{2}\sum_{i=1}^{N}\ln\left[1 + \gamma(\eta_i'\eta_i - 1)\right] - N\ln\left[1 - \Phi(-z)\right] - \frac{1}{2}Nz^2\\ &+ \sum_{i=1}^{N}\ln\left[1 - \Phi(-z^*)\right] + \frac{1}{2}\sum_{i=1}^{N} z_i^{*2}\\ &- \frac{\sigma_S^2}{2(1-\gamma)}\sum_{i=1}^{N}(y_i - x_i\beta)'(y_i - x_i\beta),\end{aligned} \tag{9.54}$$

with parameter vector $\theta = (\beta', \sigma_S^2, \gamma, \mu, \eta)'$.

9.4.3 Generating Data According to the Model

We want to generate a small data set with $N = 10$ firms and $T_i = T = 5$ years, hence a balanced panel. As parameters, we choose $\beta_0 = 1, \beta_1 = 1, \mu = 0.2, \sigma_u = 0.8, \sigma_v = 0.5, \eta = -0.2$.

```
N <- 10
M <- 5
NM <- N*M
i <- rep(1:N,each=M)  # firm identifier
m <- rep(1:M,N)       # year identifier
b0 <- 1
b1 <- 1
mu <- .2
su <- 0.8
sv <- 0.5
h <- 0.2
```

We generate the x-values using a uniform distribution. The inefficiency and the random terms are generated using normal distributions. In R, we use h for η and g for γ. We denote the T observations $t - T$ by tmT. The suffix l indicates a long vector of length NM, suffix s a short vector of length N.

```
su2 <- su^2
sv2 <- sv^2
g <- su2/(su2+sv2)
tmT <- rep(1:M-M,N)
set.seed(5)
x <- runif(NM,0,10)
us <- abs(rnorm(N,mean=mu,sd=su))
hl <- exp(-h*tmT)
u <- rep(us,each=M)*hl
v <- rnorm(NM,sd=sv)
y <- b0+b1*x+v-u
```

We draw a plot (see Fig. 9.7) of input and output and indicate the deterministic production relation by a line (left panel). The right panel shows the $N = 10$ different time trends of firm specific inefficiencies.

We write a function that returns the value of the negative log-likelihood for a given parameter vector (a).

```
#a <- astart
LL <- function(a){
  b0 <- a[1]
  b1 <- a[2]
  s2 <- a[3]
  g <- a[4]
  mu <- a[5]
  h <- a[6]
  hl <- exp(-h*tmT)
  hi2 <- as.vector(tapply(hl,i,function(z) sum(z^2)))
  e <- y-b0-x*b1
  ei2 <- as.vector(tapply(e,i,function(z) sum(z^2)))
  he <- hl*e
```

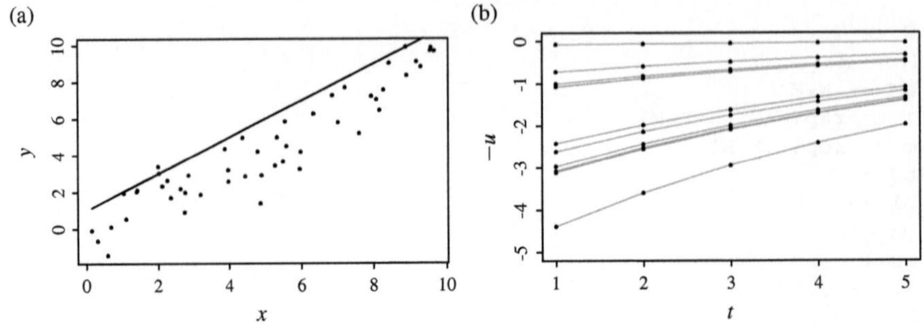

Fig. 9.7 Data and inefficiency trends. (**a**) Artificial SFA data. (**b**) Inefficiency trends

```
  hei <- as.vector(tapply(he,i,sum))
  gs2 <- g*s2
  z <- mu/sqrt(gs2)
  zisn <- g*(1-g)*s2*( 1+(hi2-1)*g );zisn
  zis <- (mu*(1-g)-g*hei)/sqrt(zisn);zis
-(  -0.5*NM*(log(2*pi)+log(s2)) -
     0.5*(NM-N)*log(1-g) -
     0.5*sum(log(1+(hi2-1)*g)) -
     N*log(1-pnorm(-z)) -
     0.5*N*z^2 +
     sum(log(1-pnorm(-zis))) +
     0.5*sum(zis^2) -
     0.5*sum(ei2)/(1-g)/s2)
}
```

We define a starting vector which we use as the argument for the log-likelihood function. Next we use `optim()` to numerically optimize the log-likelihood function.

```
astart <- c(b0,b1,su2+sv2,g,mu,h);astart

## [1] 1.0000 1.0000 0.8900 0.7191 0.2000 0.2000

LL(astart)

## [1] 46.19

o <- optim(astart*0.8,LL)
round(o$par,2)

## [1] 0.93 0.97 0.88 0.81 0.36 0.21
```

We compare our results with the results we obtain when using the function `sfa()` implemented in the frontier package. Before using the sfa-function, we load the package plm to define a dataframe for panel data.

```
library(plm)
library(frontier)
dat <- plm.data(data.frame(cbind(i,m,x,y)),c("i","m"))
reg <- sfa( y ~ x, data = dat, timeEffect = TRUE,
            truncNorm = TRUE, startVal=astart*0.8)
round(cbind(astart,o$par,coef(reg)),3)

##                astart
## (Intercept)     1.000 0.933 0.941
## x               1.000 0.966 0.967
## sigmaSq         0.890 0.883 0.679
## gamma           0.719 0.807 0.749
## mu              0.200 0.361 0.569
## time            0.200 0.213 0.213
```

We find that both functions result in very similar parameter estimates. Note that the number of firms in our generated data set is very small. Hence, the estimation is rather imprecise.

The estimates of technical inefficiency are obtained as the conditional expectation given the composed error term

$$\mathrm{E}\left[\mathrm{e}^{-U_{it}}\,|E_i\right] = \left(\frac{1-\Phi\left(\eta_{it}\sigma_i^* - \mu_i^*/\sigma_i^*\right)}{1-\Phi\left(-\mu_i^*/\sigma_i^*\right)}\right)\exp\left(-\eta_{it}\mu_i^* + \frac{1}{2}\eta_{it}^2\sigma_i^{*2}\right),$$

$$\text{with}\quad \mu_i^* = \frac{\mu\sigma_V^2 - \eta_i' E_i \sigma^2}{\sigma_V^2 + \eta_i'\eta_i\sigma^2} \quad\text{and}\quad \sigma_i^* = \frac{\sigma_V^2\sigma^2}{\sigma_V^2 + \eta_i'\eta_i\sigma^2}. \tag{9.55}$$

Given the maximum likelihood estimates of the parameters, the following transformations have to be used to calculate the inefficiency estimates

$$\sigma_U^2 = \gamma\sigma_S^2 \quad\text{and}\quad \sigma_V^2 = \sigma_S^2(1-\gamma). \tag{9.56}$$

We write a function that returns the estimates of the efficiency given the maximum likelihood parameter estimates.

```
ef <- function(a){
  su2 <- a[3]*a[4];su2
  sv2 <- a[3]*(1-a[4]);sv2
  hl <- exp(-a[6]*tmT)
  hi2 <- as.vector(tapply(hl,i,function(z) sum(z^2)))
  e <- y-a[1]-x*a[2]
  he <- hl*e
  hei <- as.vector(tapply(he,i,sum))
  mus <- (a[5]*sv2-hei*a[3])/(sv2+hi2*a[3]);mus
  ss <- sqrt((sv2*a[3])/(sv2+hi2*a[3]));ss
  musl <- rep(mus,each=M)
  ssl <- rep(ss,each=M)
  (1-pnorm(hl*ssl-musl/ssl))/(1-pnorm(-musl/ssl))*
    exp(-hl*musl+0.5*hl^2*ssl^2)
}
E.u <- ef(o$par)
```

In the frontier package, the function `efficiencies()` returns a matrix of estimated efficiencies with a sfa-object as an argument for the function.

```
E.u.sfa <- as.vector(t(efficiencies(reg)))
round(cor(E.u,E.u.sfa),3)

## [1] 1
```

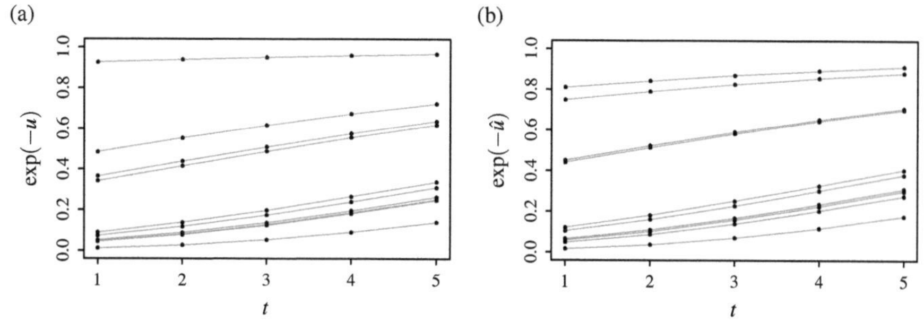

Fig. 9.8 Generated and estimated inefficiency trends. (**a**) Generated. (**b**) Estimated

These estimates are almost identical to the estimates we obtained previously. Finally, we graphically compare the generated and estimated inefficiency trends (see Fig. 9.8).

9.5 Recommended Reading

Panel data stochastic frontier models are discussed in

- Kumbhakar SC, Lovell CK (2000) Stochastic frontier analysis. Cambridge University Press, Cambridge
- Greene WH (2008) The econometric approach to efficiency analysis. In: Fried HO, Lovell CAK, Schmidt SS (eds) The measurement of productive efficiency and productivity growth, chap 2. Oxford University Press, New York, pp 92–250.

The panel data model discussed in detail in this chapter which is implemented in the frontier package

- Coelli T, Henningsen A (2013) Frontier: stochastic frontier analysis. URL http://CRAN.R-Project.org/package=frontier, r package version 1.1–0.

has been suggested by

- Battese GE, Coelli TJ (1992) Frontier production functions, technical efficiency and panel data: with application to paddy farmers in India. J Prod Anal 3:153–169.

9.6 Exercise

We use data which have been analyzed by Battese and Coelli (1992; 1995) and are available in the frontier package. The data set contains data for 43 farms observed over a period of 8 years for output, area, labor, and fertilizer. Read the data into a panel-dataframe using the following commands:

```
library(frontier)
data(riceProdPhil)
d <- subset(riceProdPhil,
            select=c(FMERCODE,YEARDUM,PROD,AREA,LABOR,NPK))
names(d) <- c("firm","year","y","a","l","f")
pd <- plm.data(d,c("firm","year"))
```

1. Use the function `summary()` for an overview of the data.
2. Use the data for year 8 and plot the three bivariate scatterplots with one input and output. Plot the data again in logs.
3. Estimate a simple cross-sectional production function of the Cobb–Douglas type without restricting the elasticities to sum to unity using data from year 8 only.
4. Use the function `lm()` and the panel data set to obtain time invariant efficiency estimates for the $N = 43$ firms using a fixed effects model. Display the estimated inefficiencies graphically.
5. Estimate a fixed effects model allowing the firm specific inefficiency to vary in time according to a simple one-parameter function using the function `lm()`. Display the estimated inefficiencies graphically.
6. Estimate the SFA model suggested by Battese and Coelli (1992) with time varying inefficiencies using the function `sfa()` in the frontier package. Display the estimated inefficiencies graphically.

References

Aigner D, Lovell CK, Schmidt P (1977) Formulation and estimation of stochastic frontier production function models. J Econ 6:21–37

Battese GE, Coelli TJ (1992) Frontier production functions, technical efficiency and panel data: with application to paddy farmers in India. J Prod Anal 3:153–169

Battese GE, Coelli TJ (1995) A model for technical inefficiency effects in a stochastic frontier production function for panel data. Empir Econ 20:325–332

Coelli T, Henningsen A (2013) Frontier: stochastic frontier analysis. URL http://CRAN.R-Project.org/package=frontier, r package version 1.1–0

Greene W (2005) Fixed and random effects in stochastic frontier models. J Prod Anal 23(1):7–32, URL http://dx.doi.org/10.1007/s11123-004-8545-1

Greene WH (2008) The econometric approach to efficiency analysis. In: Fried HO, Lovell CAK, Schmidt SS (eds) The measurement of productive efficiency and productivity growth, chap 2. Oxford University Press, New York, pp 92–250

Kumbhakar SC, Lovell CK (2000) Stochastic frontier analysis. Cambridge University Press, Cambridge

Meeusen W, van de Broeck J (1977) Efficiency estimation from Cobb–Douglas production functions with composed errors. Int Econ Rev 18:435–444

Pitt MM, Lee LF (1981) The measurement and sources of technical inefficiency in the Indonesian weaving industry. J Dev Econ 9:43–64

Schmidt P, Sickles RC (1984) Production frontiers and panel data. J Bus Econ Stat 2(4):367–374

Functions Index

anova, 77
apply, 3
as.factor, 97
as.matrix, 42
as.vector, 42

barplot, 46
Benchmarking package, 167
between, 100
boot package, 130, 153
boot.sw98, 180

C, 1
car package, 15
cbind, 13
choose, 162
colMeans, 177
colSums, 45
combn, 162

data.frame, 2
dea, 148, 154, 157, 167, 177, 179
dea.plot, 152
dea.plot.isoquant, 154
density, 179
dev.copy, 9
diag, 6
dnorm, 7

efficiencies, 220

FEAR package, 180
for, 3
frontier package, 219

help, 2
hist, 8

linearHypothesis(), 14
lm, 97, 190, 206

matrix, 3
maxLik package, 68
mean, 1
median, 30
miCecon package, 78

names, 10
ncol, 13
nrow, 13

optim, 18, 26, 29, 67, 196, 219
optimize, 16

pdata.frame, 100, 207
persp, 21
plm, 100, 207
plm.data, 219
plm package, 99, 207
plot, 8
pnorm, 7

qnorm, 7
quantile, 30

rbind, 14
read.csv2, 6
read.table, 6
rep, 3
rnorm, 7

sample, 177
scatterplot3d package, 61
seq, 3
set.seed, 7

A. Behr, *Production and Efficiency Analysis with R*,
DOI 10.1007/978-3-319-20502-1

sfa, 219
simplex, 130, 153, 155, 156
solve, 5, 40
summary, 9

tapply, 3
translogEst, 78

unlist, 98

vcov, 10

which, 16, 190
Within, 100
write.csv2, 6
write.table, 6

Subject Index

Analysis of variance, 77
Anova, 77

Between estimator, 86, 96, 101
Bootstrap, 164, 171

CCR-model, 141
CD, 59
CES, 62
  approximation, 78
  elasticity of substitution, 65
  estimation, 66
  properties, 63
Charnes–Cooper–Rhodes-model, 141
Cobb–Douglas, 59, 101
  elasticity of substitution, 60
  estimation, 60
  properties, 59
Cols, 189
  with R, 189
Combined error term, 184
Confidence intervals, 180
Constant elasticity of substitution, 62
Constant prices, 44
Convex sets, 118
Corrected ols, 189
  with R, 189
Cost minimization, 113

Data, 2
Data envelopment analysis, 141
  stochastic, 161
  with R, 151
Dea, 141
  bootstrap, 171
  sampling, 167
  stochastic, 161
Demeaning, 98
Dual, 126, 129, 130
Duality, 125
Duality theorems, 126

Efficiency, 135
  time invariant, 204
  time trend, 209
Efficient frontier, 136
Elasticity of substitution
  Cobb–Douglas, 60
Elasticity of substitutions
  CES, 65
Employment inverse, 37
Employment model, 36
Employment multipliers, 37, 54
Envelopment form, 145, 155
Error components, 85, 87
Error term
  sfa, 184
Euler's theorem, 60, 64
Exponential distribution, 15

Farrell-efficiency, 147
Final demand, 35
Final domestic use, 44
Fisher information, 19
Fixed effects, 94
Fixed effects estimator, 86
Frontier, 136

Generalized least square, 90
GLS estimator, 90
Graphics, 8
  export, 8

Hessian matrix, 21
Histogram, 8

Idempotent matrix, 89
Indirect inputs, 35

A. Behr, *Production and Efficiency Analysis with R*,
DOI 10.1007/978-3-319-20502-1

Induced production, 54
Inefficiency
time varying, 208, 211, 212
Input, 33
Input coefficients, 33
Input–output analysis, 41
Input–output table, 41
Input-oriented model, 149
Inputs
direct, 35
indirect, 35
Inverse, 5
Iso-cost line, 114
Iso-profit line, 116

Labor coefficients, 36
Labor productivity, 49
Leontief-inverse, 35
Likelihood, 16
Linar program, 117
Linear model, 9
Linear production model, 33
Linear programming, 113
Linear regression, 9
Linear restrictions, 10, 62
Linearization, 60
Log-likelihood, 16, 92
sfa, 192
Loops, 3

Matrix, 4
Maximization, 116, 120
Maximum likelihood, 15
idea, 15
properties, 18
sfa, 191
Maximum likelihood estimator, 91
Minimization, 113, 122
Multiplier form, 145, 153
Multipliers, 37

Newton–Raphson algorithm, 23
Normal half-normal model, 203
Numerical optimization, 18, 22
CES, 70

Objective function, 114, 116
Optimization, 18, 22
Output, 33
Output-oriented model, 148

Panel data, 83
between estimator, 96, 101
Cobb–Douglas, 101
demeaning, 98
efficiency, 204, 208
error components, 85
estimators, 85
fixed effects, 94, 96, 101
generation, 95
ml estimator, 91
pooled estimator, 96, 101
random effects, 87, 94, 101
sfa, 203
two-ways effects, 101
within estimator, 96, 101
Panel data sfa, 203
data generation, 217
efficiency scores, 220
estimation, 219, 220
log-likelihood, 214, 218
model, 212
Panel dataframe, 100
Pareto–Koopman-efficiency, 147
Perspective plot, 21
Plot, 8
Pooled estimator, 85, 96, 101
Population and samples, 162
Price index, 44
Primal, 126, 129, 131
Primal-dual correspondence, 125, 131
Production correspondence, 144
Production function, 57
CES, 62
Cobb–Douglas, 59
transcendental logarithmic, 71
Production matrix, 34
Production model, 33
numerical example, 38
Productivity, 135
Productivity of labor, 49
Profit maximization, 116

Random effects, 87, 94, 101
log-likelihood, 92
Random effects estimator, 90
Rao–Cramer-lower bound, 19
Reference set, 138
Regression, 9
Replication, 3
Resampling, 164
Restrictions, 10

Scatterplot, 8
Sectors, 46
share in output, 46
share in value added, 46
Separability, 74

Sequence, 3
Sfa, 183
    data generation, 185
    efficiency scores, 195, 198
    error term, 184
    inefficiency, 193
    log-likelihood, 192
    maximum likelihood, 191
    ml estimation, 195
    normal-half normal model, 185
    panel data, 203
Simar–Wilson
    simplified algorithm, 176
Simar–Wilson algorithm, 178
Simar–Wilson approach, 176
Simplex algorithm, 120
Simplex tableau, 120, 127, 129
    solving, 129
Slack variables, 119
Stochastic dea, 161, 176
    algorithm, 178
    confidence intervals, 180
Stochastic frontier analysis, 183

Taylor series, 23
Transcdendental logarithmic, 71
Transcendental logarithmic
    estimation, 75
    separability, 74
Transcental logarithmic
    properties, 72
Translog, 71
    CES approximation, 78
    estimation, 75
    properties, 72
    separability, 74
Two-ways effects, 101

Weights, 140
Within estimator, 86, 96, 101

MIX
Papier aus verantwortungsvollen Quellen
Paper from responsible sources
FSC® C105338

If you have any concerns about our products,
you can contact us on
ProductSafety@springernature.com

In case Publisher is established outside the EU,
the EU authorized representative is:
Springer Nature Customer Service Center GmbH
Europaplatz 3, 69115 Heidelberg, Germany

Printed by Libri Plureos GmbH
in Hamburg, Germany